IDENTIFICATION GUIDE

ALLIED FIGHTER AIRCRAFT 1939–45

AMERICAN, BRITISH, FRENCH, SOVIET, DUTCH, POLISH

IDENTIFICATION GUIDE

ALLIED FIGHTER AIRCRAFT 1939–45

AMERICAN, BRITISH, FRENCH, SOVIET, DUTCH, POLISH

CHRIS CHANT

amber BOOKS

This edition first published in 2022

Published by
Amber Books Ltd
United House
London N7 9DP
United Kingdom
www.amberbooks.co.uk
Instagram: amberbooksltd
Facebook: amberbooks
Twitter: @amberbooks

Project Editor: Michael Spilling
Designer: Hawes Design
Picture research: Terry Forshaw

ISBN: 978-1-83886-193-3

Printed in Thailand

PHOTOGRAPH AND ILLUSTRATION CREDITS

All photographs courtesy of Art-Tech/Aerospace except for 10 and 172 courtesy of Cody Images.All artworks courtesy of Art-Tech/Aerospace except for 12, 13(b), 14(b), 21(b), 22, 24(t), 28–29(all), 30(t), 33(b), 34(b), 36(both), 48(t), 51(t), 52, 80(c), 81(t), 83(t), 95, 104, 106(c), 116(b), 144, 154(b), 165(t), 168, 174, 176–179(all) courtesy of Alcaniz Freson's S.A.

Contents

Introduction

The fighter is essentially a defensive weapon whose capabilities can also be used offensively. In the Battle of Britain, for example, the Royal Air Force's fighters were used purely defensively to protect the United Kingdom, either by destroying the German bombers or tackling the German escort fighters to allow other RAF fighters an easier approach to the German bombers. Modified by time and occasion, the same basic tenet held true throughout the war. As in World War I, therefore, the fighter of World War II was an arbiter of the air war as it either aided or prevented other warplanes from going about the tasks that could affect the course of the war in direct terms – such as bombing industrial targets and supporting ground offensives.

Hurricanes in 'vic' formation

This photograph of some uncoded, newly-delivered Hawker Hurricane Mk Is shows two sets of three aircraft in the classic 'vic' formation used by the RAF up until 1941, before the British adopted the more effective 'finger four' based on the German *Schwarm*.

Polikarpov I-16s in flight

Although obsolete in 1941 when the Germans invaded the Soviet Union, the Polikarpov I-16 proved to be a tough and manoeuvrable aircraft that gave a good account of itself against the superior German Bf 109.

The basics of fighter tactics were unaltered in concept from those developed in World War I, and were therefore derived from speed, climb rate, overall agility with emphasis on turn rate, armament and the pilot's reaction time and exploitation of sun and cloud. Surprise remained the trump card. The major advance between the wars was the advent of voice communication by radio, which allowed pilots to pass on and receive information and orders, and from a time very late in the 1930s to receive instruction from ground controllers watching the development of the air battle on radar.

In September 1939, Royal Air Force (RAF) Fighter Command's tactics were derived from the concept that the air threat to the UK would comprise German bombers, flying in close formation and unescorted by fighters as the latter were unable to reach Britain's shores from airfields in Germany. It was believed that the era of the dogfight had passed, and thus rigid air fighting tactics were introduced. By means of complex and lengthy manoeuvres, these were designed to bring the maximum number of guns to bear on the bomber formation. The fighting unit was thus the tight 'vic' (V formation) of three fighters.

As events very rapidly proved, such formulaic attacks were of no use. The *Luftwaffe*'s fighter arm was decidedly superior as its pilots had gained invaluable tactical experience under real combat conditions during the Spanish Civil War (1936–39), and had devised and proved the ideal fighter formation, the *Rotte*. This comprised two fighters, spaced some 180m (600ft) apart, and the primary task of wingman was to guard his leader from an attack from the beam or the rear as the leader made the tactical decisions and also covered his wingman.

The four-fighter *Schwarm* was the logical expansion of the *Rotte*, with two pairs of aircraft rather than just two aircraft. When, in 1941, the RAF eventually copied the *Luftwaffe* and adopted this pattern, it was called the 'finger-four', as the relative positions of the fighters was similar to the tips of the finger.

Battle of Britain

Despite the lessons of the brief air fighting during Germany's attack on the Low Countries and France, in June 1940 the British entered the Battle of Britain with the unwieldy 'vic'. The *Luftwaffe* failed to exploit opportunities, such as the RAF's radar gap at low level, however, and as the battle continued and grew more intense, more able tactical leaders adopted the looser type of formation which improved manoeuvrability without sacrificing cover. This meant that the British fighters were steadily better placed to engage the German fighters as the latter were switched, quite wrongly, from free-roaming hunts for British fighters to close escort of the bombers, so losing the advantage of the fighters' agility and surprise.

The British also evolved different tactics against the bombers. In one popular approach, the fighters, flying line abreast, flew head-on at the bombers to minimize the time in which the fighter escort could intervene, and also often to break up the bomber formation for piecemeal engagement by the fighters. In another, based on the so-called 'big wing' concept, a large formation was grouped to make a simultaneous attack more effective than the same number of aircraft arriving separately. The Duxford Wing, of three and later five squadrons, was created but not notably successful as its formation in the air almost always took too long.

After the battle of Britain, Fighter Command tried to retain the initiative by luring the *Luftwaffe* to battle over France with 'Circus' operations, but this failed to gain the sought-for advantage as the Germans generally refused to rise to the bait.

Soviet air tactics

The Soviet air arm still used the three-aircraft 'vic' at the start of the German invasion of the Soviet Union in June 1941, but the pilots who survived the mass destruction of the Soviet air units in the first weeks of the campaign then started to copy the German fighter tactics in the form of eight fighters in two loose sections of four. This improved matters, but the efforts of the Soviet fighter arm were still severely hampered by the lack of ground-control radar, the general absence of radio (forcing pilots to communicate visually) and the lack of adequate air gunnery training, which meant that fire was all too often inaccurate and opened at too great a range. The training of pilots had always emphasised adherence to group attacks rather than the exercise of small unit or individual initiative, and the inevitable fragmentation of units in combat had a decidedly adverse effect on Soviet pilots' ability to function effectively.

Generally, the pilots were reluctant to engage the German bombers closely, and often ended their attacks too quickly before inflicting decisive damage. When engaged by German fighters, the Soviet pilots adopted the defensive circle manoeuvre learned in the Spanish Civil War, in which each fighter covered that ahead of it, before seeking to break away at low level.

However, by late 1943, the quality of Soviet fighter pilots and their equipment had improved radically, and, in combination with their huge numerical superiority, they gradually gained total command of the air over the Eastern Front.

P-51D 'Finger Four'

In this photograph from 1944, four USAAF P-51D Mustangs from the 308th Fighter Squadron fly in the classic 'finger four' formation used by most Allied air forces from 1942 onwards.

US fighter tactics

Although the United States Army Air Force's (USAAF) fighter tactics from the involvement of the United States in World War II from December 1941 were still too rigid, moves toward the 'finger four' had already begun, and spread rapidly through the USAAF and US Navy. The major innovation of the concept was the 'Thach weave' developed by Lieutenant Commander John Thach of the US Navy in a largely successful effort to give the Grumman F4F Wildcat a better chance against the technically superior Mitsubishi A6M 'Zero'. The new tactic was based on two pairs of fighters abreast of each other with the pairs about 365m (1200ft) apart. When an attack was made on one pair, the other broke towards it to engage, and the pair being attacked broke towards the pair protecting it. This created a scissors effect as the two pairs crossed, and meant that the pursuing Japanese fighter(s) were faced with a beam attack if they held their course, a head-on attack if they turned towards the attackers, or a rear attack if they turned away.

Ultimately the US advantage in training, experience and aircraft capability meant that American pilots were able to meet the Axis pilots on more than equal terms and decimate their opponents, who were, from 1944, decidedly inferior in experience and numbers.

Chapter 1

France

By 1939, the French Air Force was beginning to emerge from the political and technical doldrums in which it had languished since World War I. New thinking and new aircraft were evident, but the problem the *Armée de l'Air* now had to face was whether it was too little or too late, or worst of all, both too little and too late.
The fighter arm had at last disposed of nearly all of its obsolete first-generation monoplane fighters, but was still reliant on the indifferent Morane-Saulnier MS.406, pending the arrival of more advanced fighters such as the Dewoitine D.520.

First-generation monoplane fighter
The Morane-Saulnier MS.406C.1 was created on the basis of several modern features, but was at best an indifferent fighter with performance decidedly inferior to that of the German fighters it would have to face.

The *Armée de l'Air*

1939–40

The *Armée de l'Air* was numerically large and appeared formidable, but was in fact dependent largely on obsolescent aircraft, and was still suffering from the effects of political antipathies and the nationalization of the French aero industry in the mid-1930s.

By 1937, it was clear that the French air force needed more modern aircraft to replace inferior fighters such as the Dewoitine D.500 and D.501, which were still in service with fighter squadrons including the famous *Cigognes* (storks). This squadron, which was part of *Groupe de Chasse* (GC) I, was initially stationed at Chartres-Champbol. However, only a matter of days before Germany triggered World War II by invading Poland on 1 September 1939, it moved to Beauvais-Tillé after replacing its D.500 and D.501 fighters with Morane-Saulnier MS.406 machines (each armed with one 20mm/0.79in cannon). These were among the most modern fighters on the strength of the *Armée de l'Air* at the start of World War II. France's programme to manufacture more than 2500 modern aircraft, including Bloch MB.170 bombers as well as Dewoitine D.520 fighters, was a direct response to the difficulties in which the government had found itself through a comment apparently made by the commander-in-chief of the air force. He said that less than half of France's total of about 1400 first-line aircraft would be ready for action at a moment's notice, and that most of these machines were also

ORGANIZATION

l'Armée de l'Air
Single-seat fighter strength, 1 September 1939

MS.405 & 406 (573)	MB.151 &152 (120)	Curtiss Hawk 75 (172)	Caudron C.714 (7)

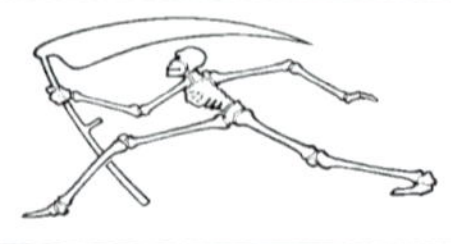

The emblem of the *3e Escadrille, GCII/1* was this depiction of death as a long-striding skeleton carrying a wide-bladed scythe.

Specifications

Crew: 1
Powerplant: 641kW (860hp) Hispano-Suiza liquid-cooled V-12
Maximum speed: 402km/h (250mph)
Range: 700km (435 miles)
Service ceiling: 11,000m (36,090ft)
Dimensions: span 12.09m (39ft 8.25in); length 7.94m (26ft 0.5in); height 2.42m (7ft 11.25in)
Weight: 1929kg (4244lb) loaded
Armament: 1 x 20mm (0.79in) HS9 cannon and 2 x 7.5mm (0.295in) MAC1934 MGs

▲ Dewoitine D.510C.1

3e Escadrille / Groupe de Chasse II/1 (GC IV/1), Etampes, 1938

The D.510 was an interim monoplane fighter with fixed landing gear and open accommodation, but moderately good armament in the form of one 20mm (0.79in) fixed forward-firing cannon (between the cylinder banks to fire through the hollow propeller shaft) and two 7.5mm (0.295in) machine guns.

FIGHTER AIRCRAFT: DEWOITINE D.520C.1

Unit	Operational	Victories	Losses
GC I/3	17/04/40	50 +18 prob	32
GC II/3	10/05/40	31 +15 prob	20
GC II/7	20/05/40	12 + 4 prob	14
GC III/3	28/05/40	8 + 2 prob	17
GC III/6	10/06/40	7 + 0 prob	2

obsolescent. There was a large element of truth in this, but it also reflected the political pressure to which service commanders customarily resorted to secure resources in the face of demands by the other two service commanders.

And while resources were one matter, another was the air force's lack of clear thought about how it should best be used. Even within the service there were several schools of operational and tactical

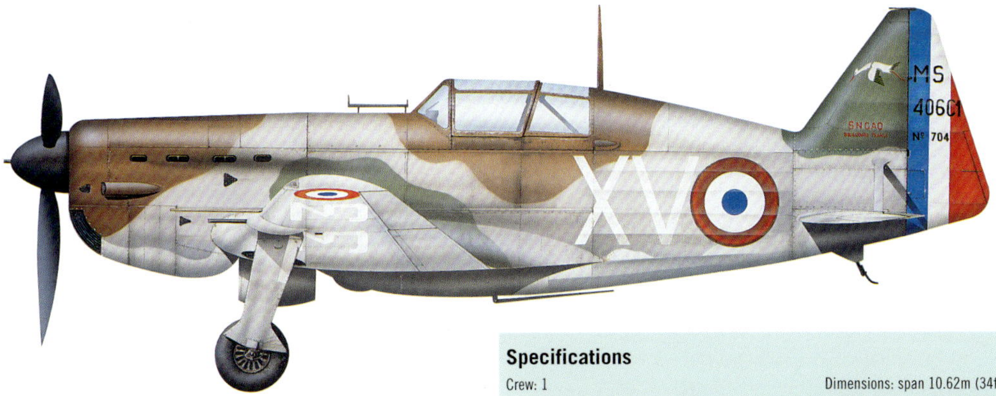

▲ Morane Saulnier MS.406C.1

1e Escadrille / GC I/2 / Nimes, July 1940

The MS.406 was France's first 'modern' monoplane fighter, but was in all major respects, except firepower, an indifferent warplane with little to commend it except ease of manufacture and availability.

Specifications

Crew: 1

Powerplant: 641kW (860hp) Hispano-Suiza liquid-cooled V-12

Maximum speed: 490km/h (304mph)

Range: 750km (466 miles)

Service ceiling: 9400m (30,840ft)

Dimensions: span 10.62m (34ft 9.5in); length 8.17m (26ft 9.33in); height 3.25m (10ft 8in)

Weight: 2471kg (5448lb) loaded

Armament: 1 x 20mm (0.79in) HS9 or HS404 cannon and 2 x 7.5mm (0.295in) MAC1934 MGs

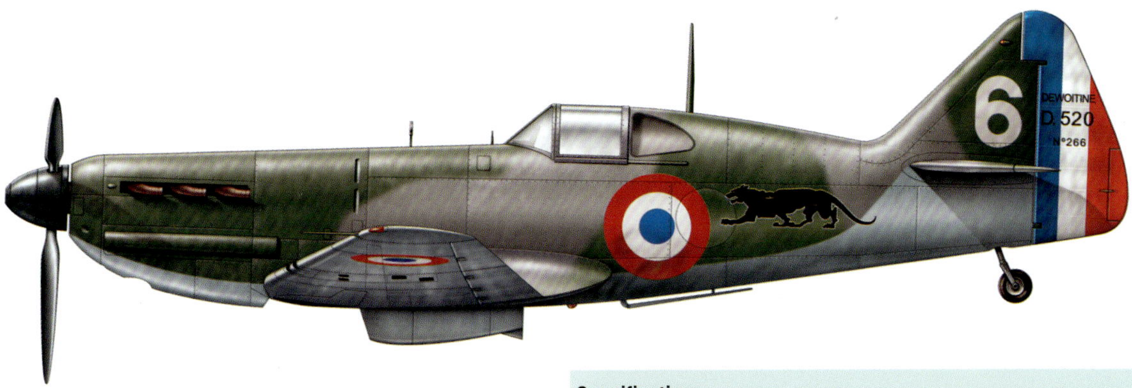

▲ Dewoitine D.520C.1

GC II/7 / France, July 1940

The D.520 was without doubt the best single-seat fighter available to the French in the first part of World War II, but was at that time delivered in only small numbers, and its pilots could therefore exercise no real impact on events.

Specifications

Crew: 1

Powerplant: 686kW (920hp) Hispano-Suiza 12Y-45 liquid-cooled V-12

Maximum speed: 535km/h (332mph)

Range: 900km (553 miles)

Service ceiling: 11,000m (36,090ft)

Dimensions: span 10.20m (33ft 5.5in); length 8.76m (28ft 8.75in); height 2.57m (8ft 5.25in)

Weight: 2783kg (6134lb) loaded

Armament: 1 x 20mm (0.79in) Hispano-Suiza HS-404 cannon and 4 x 7.5mm (0.295in) MAC1934 MGs

The Armée de l'Air Groupe de Chasse

The French core fighter unit in 1939/40 was the *groupe de chasse* (wing), which comprised a number of *escadrilles* (squadrons) sometimes on different bases and, in time of war, themselves allocated to various *groupements* (groups) with other types of aircraft. Thus on 10 May 1940 *Groupement de Chasse 21* in the *Zone d'Operations Aeriennes Nord,* comprised GC I/1 at Chantilly-les-Aigles, GC II/1 at Buc, GC III/3 at Beauvais-Tille, GC II/10 at Rouen-Boos and GC III/10 at Le Havre-Octeville.

1e Escadrille

2e Escadrille

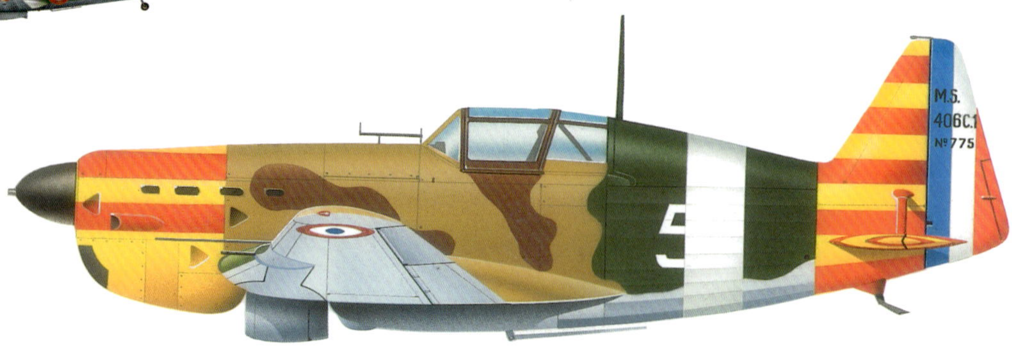

Specifications

Crew: 1

Powerplant: 641kW (860hp) Hispano-Suiza 12Y-31 liquid-cooled V-12

Maximum speed: 490km/h (304mph)

Range: 750km (466 miles)

Service ceiling: 9400m (30,840ft)

Dimensions: span 10.62m (34ft 9.5in); length 8.17m (26ft 9.33in); height 3.25m (10ft 8in)

Weight: 2471kg (5448lb) loaded

Armament: 1 x 20mm (0.79in) HS-9 or HS-404 cannon and 2 x 7.5mm (0.295-in) MAC1934 MGs

Morane-Saulnier MS.406C.1

Escadron de Entrainment / Toulouse, 1941

The MS.406C.1 was numerically the most important French fighter in the campaign that led to France's defeat in May–June 1940. Production amounted to 1077 aircraft, and the survivors were used mostly for training from 1941.

thought, and this led to interminable argument and delay. All this wrangling continued even as Germany, France's eastern neighbour on the other side of the great Rhine river, was steadily building up her forces on the basis of a fairly concentrated industrial effort, and within the context of clear military imperatives and only limited inter-service rivalry.

Operational deficiencies

The combined effect of France's inadequate aeronautical programmes and the indecision evident at all levels, from the government right down to the middle echelons of the armed forces, inevitably meant that the French Air Force was materially and operationally deficient relative to the air force it would have to face in combat after September 1939. The problem was exacerbated by the fact that while the French air force had no experience of air warfare under modern conditions, the German air force certainly did. The commitment of the Legion Condor to the Spanish Civil War (1936–39) had brought it that considerable advantage. Fighting on the side of the victorious Nationalists, German fighters and their pilots had been tried against modern Soviet fighters flying on behalf of the Republican side.

Germany's fighters were therefore combat-proven and in the process of steady development, while their pilots had worked out a system of tactics that were admirably suited to the demands of high-speed air combat. The Germans had also developed a high level of skill in the cooperation of the various elements of their air force, and in the flying of independent bombing campaigns and the means whereby the air force could best provide tactical air support for the ground forces.

When it became evident in the mid-1930s that another major European war was probably if not actually certain, France sought to respond through an intense programme of re-equiping and modernization in 1938–39. However, this was an effort that was taken much too late to be meaningful, and much too inadequately to be effective.

▲ First flight

This photograph shows pilot Marcel Doret taking a Dewoitine D.520 on its first flight in October 1938.

Specifications

Crew: 1
Powerplant: 768kW (1030hp) Rolls-Royce Merlin III liquid-cooled V-12
Maximum speed: 582km/h (362mph)
Range: 636km (395 miles)
Service ceiling: 9725m (31,900ft)
Dimensions: span 11.22m (36ft 10in); length 9.11m (29ft 11in); height 2.69m (8ft 10in)
Weight: 2624kg (5784lb) loaded
Armament: 8 x 7.7mm (0.303in) Browning MGs

▲ Supermarine Spitfire Mk I

Experimental and development service, France 1940

The French Air Force did not use the British-designed Spitfire as an operational fighter, but received one aeroplane in 1938 for experimental and development purposes. It is believed that this aeroplane was captured by the Germans late in the Battle of France, during June 1940.

The defeat of France

May–June 1940

Woefully ill-prepared, and with only small British forces available to aid her, France lacked the strength and organization to halt, let alone defeat, the German ground and air onslaught that fell on her on 10 May 1940. Less than seven weeks later, she was compelled to surrender.

WHEN THE WAR STARTED, the *Armée de l'Air* was in a very poor state, and could muster only 826 combat-ready fighters. More aircraft had been built, but these lacked essential equipment or final preparation: some lacked gun sights or propellers, and others had guns that had not been harmonized.

Thus when the German offensive started in May 1940, the Germans could deploy not only more aircraft but also more experienced pilots, many of them veterans of the Spanish Civil War. The Germans advanced swiftly through France and both out-thought and out-fought all Allied opposition. The French and their British allies struggled hard over France, but could achieve nothing significant to stem the German tide. In the air, the French fighter arm was at times able to achieve local successes, but these were of little use when the air support squadrons and their covering fighters were often out

ORGANIZATION

l'Armée de l'Air
May 1940

- Zone d'Operations Aeriennes Nord
- Zone d'Operations Aeriennes Est
- Zone d'Operations Aeriennes Sud
- Zone d'Operations Aeriennes des Alpes

FIGHTER AIRCRAFT: ZOAN (GROUPEMENT DE CHASSE 21)

Unit	Type	Base	Serviceable
GC I/1	MB.152	Chantilly	n/a
GC II/1	MB.152	Buc	n/a
GC II/10	MB.151/152	Rouen	n/a
GG III/10	MB.151/152	Le Havre	n/a

Specifications

Crew: 1
Powerplant: 746kW (1000hp) Gnome-Rhone 14N-25 air-cooled 14-cylinder radial
Maximum speed: 509km/h (316mph)
Range: 540km (335 miles)
Service ceiling: not available
Dimensions: span 10.54m (34ft 7in); length 9.10m (29ft 10.25in); height 3.03m (9ft 11.33in)
Weight: 2800kg (6160lb) loaded
Armament: 2 x 20mm (0.79in) Hispano-Suiza HS-404 cannon and four 7.5mm (0.295in) MAC1934 MGs

▲ Bloch MB.152C.1

3e Escadrille / GC II/1 / Buc, May 1940

This unit was part of the *Groupement de Chasse 21* facing the main German offensive. Of some 300 such aircraft delivered by January 1940, about two-thirds were non-operational for lack of the required propeller.

of contact with the French Army units they were tasked to support, in part due to the poor coordination of communications between the army and the air force, and in part to the army's obsolescent and unreliable radio equipment.

As it became clear that the campaign was lost, the French high command ordered what remained of the *Armée de l'Air* to French colonies in North Africa – in order, it believed at the time, that the fight could be continued. Yet the Vichy government, which became the official German-approved power in occupied France after the armistice, now ordered the dissolution of many of the air force squadrons. During the Battle of France, the French lost more than 750 aircraft, and the Germans more than 850 – the French fighters did inflict significant casualties.

▲ Dewoitine D.520C.1

4e Escadrille / Groupe de Chasse II/7 / Tunisia, 1942 (Vichy French air force)

The D.520 was one of the modern warplanes that Vichy France was allowed to keep and, indeed, to manufacture. In the Battle of France, the D.520 was credited with shooting down 114 German aircraft for the loss of 83 of its own number.

Specifications

Crew: 1
Powerplant: 686kW (920hp) Hispano-Suiza 12Y-45 liquid-cooled V-12
Maximum speed: 535km/h (332mph)
Range: 900km (553 miles)
Service ceiling: 11,000m (36,090ft)
Dimensions: span 10.20m (33ft 5.5in); length 8.76m (28ft 8.75in); height 2.57m (8ft 5.25in)
Weight: 2783kg (6134lb) loaded
Armament: 1 x 20mm (0.79in) Hispano-Suiza HS-404 cannon and 4 x 7.5mm (0.295in) MAC1934 MGs

▲ Bell P-39N Airacobra

GC II/6 'Travail' / Free French Air Force, 1943

The American-made Airacobra was one of several types of Allied warplane used by the French Air Force later in World War II. With its sturdy construction, good protection and heavy armament, the P-39 was an effective ground-attack fighter.

Specifications

Crew: 1
Powerplant: 895kW (1200hp) Allison V-1710-85 liquid-cooled V-12
Maximum speed: 605km/h (376mph)
Range: 483km (300 miles)
Service ceiling: 11,665m (38,270ft)
Dimensions: span 10.36m (34ft 0in); length 9.19m (30ft 2in); height 3.79m (12ft 5in)
Weight: 3992kg (8800lb) loaded
Armament: 1 x 37mm (1.46in) M4 cannon, 2 x 12.7mm (0.5in) Browning MGs, 4 x 7.62mm (0.3in) Browning MGs, and 227kg (500lb) of bombs

Chapter 2

United Kingdom and Commonwealth

The UK started World War II with a fighter arm that was comparatively small and inexperienced, but had good aircraft in its Hawker Hurricane and Supermarine Spitfire fighters. Over the next six-and-a-half years, this fighter arm grew greatly in size and capability through its successes in the Battle of Britain and the campaigns that followed in Burma, North Africa, Italy and, finally, northwest Europe. The Spitfire remained in steady development and was supplemented by last-generation piston-engined fighters as well as the Gloster Meteor, the only jet-powered Allied fighter to see service in World War II.

Supermarine Spitfire Mk XII

The first production variant of the Spitfire with a Griffon engine, the clipped-wing Spitfire Mk XII entered service in 1943 and was built to the extent of 100 aircraft. The engine drove a four-blade propeller.

RAF Fighter Command

1939–41

Fighter Command was dedicated to the air defence of the UK. It was growing rapidly in the years before the war, and provided fighters for the French campaign. It was Fighter Command that prevented the *Luftwaffe* from winning the Battle of Britain, and then went on the offensive.

FIGHTER COMMAND was created in 1936 in recognition of the fact that as the RAF expanded during the rearmament phase before World War II, more specialized applications were needed for fighter, bomber and coastal aircraft. On 20 May 1926, Fighter Command had been established as a group within the Inland Area, and on 1 June 1926 this Fighting Area, as it was then called, was transferred to the Air Defence of Great Britain. The Fighting Area was raised to command status in 1932 and renamed on 1 May 1936.

In the following years, the command was expanded greatly and its obsolescent biplanes were replaced by two of the most celebrated warplanes ever to fly with the RAF, namely the Hawker Hurricane and Supermarine Spitfire monoplanes. Fighter Command's decisive moment arrived in the summer of 1940 as the *Luftwaffe* tried to gain air supremacy over the English and UK as a prerequisite to the launch of a seaborne invasion.

Fighter Command was divided into a number of groups, each allocated its own area of the UK. No. 11 Group covered the main weight of the German attack, as it operated over south-east England and London. It was reinforced by No. 10 Group, which covered south-west England, and No. 12 Group, which covered the Midlands and the north of England. In the end, the Germans failed to win air superiority, although the RAF had eaten very deep into its reserves of pilots by the middle of the battle. Aircraft shortages were never a major problem, but the numbers and physical/psychological state of their pilots was of great concern. Pilots were being killed or severely wounded at a rate greater than they could be replaced from training schools. It took Fighter Command some months to recover from its losses in the Battle of Britain and go over to the offensive.

As 1941 began, Fighter Command began the task of winning air superiority over north-western France. By May, the British squadrons began to operate as

ORGANIZATION

Royal Air Force (Marshal of the RAF Sir Cyril Newall)
September 1939

Fighter Command (Air Chief Marshal Sir Hugh Dowding)	Bomber Command (Air Chief Marshal Sir Edgar Ludlow-Hewitt)	Coastal Command (Air Chief Marshal Sir Frederick Bowhill)

RAF COMBAT SQUADRONS (SEPTEMBER 1939)

Type	UK	NW Europe	Mediterranean
Fighter	41	4	6
Heavy/medium bomber	40	2	0
Light bomber	12	6	8
Torpedo bomber	1	0	1
Flying boat	4	0	1

▲ Scramble

These Hurricanes from No. 87 Squadron were committed to the Air Component of the BEF during the Battle of France. This photograph was taken somewhere in France, March 1940.

wings under the control of a experienced wing leader. Several types of short-penetration operations were tried in an effort to draw the *Luftwaffe* into an attritional campaign and keep large numbers of fighters pinned in France, particularly after the German invasion of the Soviet Union in June 1941. Large numbers of Spitfires were despatched with small numbers of medium bombers in often vain attempts to lure the German fighters into combat. The results of these efforts were decidedly mixed, and the Germans were able to leave a mere two experienced *Jagdgeschwadern* (180 aircraft at most) in Western Europe.

Most of the tactical factors that had allowed Fighter Command to win the Battle of Britain were now reversed, and 1941 saw Fighter Command claim some 731 *Luftwaffe* fighters (although in reality only 236 were lost from all causes) for the loss of about 530 of its own fighters.

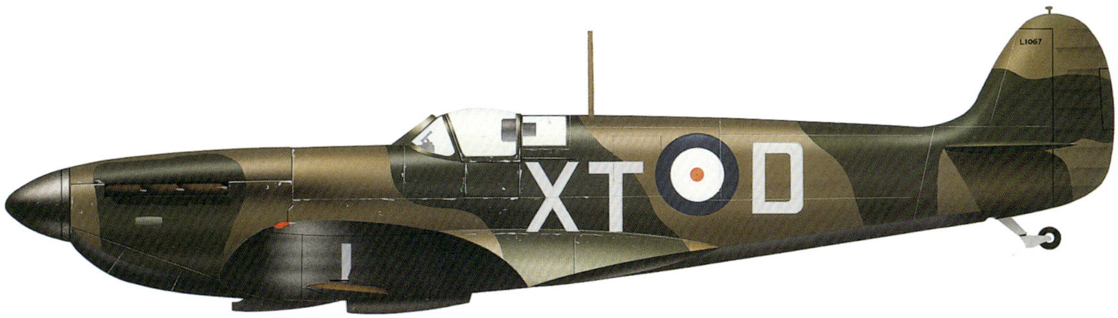

▲ Supermarine Spitfire Mk IA

No. 603 'City of Edinburgh' Squadron

Within two weeks of the outbreak of war in September 1939, the squadron began to receive Spitfires. It was operational with Spitfires in time to intercept the first German air raid on the British Isles on 16 October, when it destroyed the first enemy aircraft to be shot down over Britain in World War II.

Specifications

Crew: 1
Powerplant: 768kW (1030hp) Rolls-Royce Merlin III liquid-cooled V-12
Maximum speed: 582km/h (362mph)
Range: 636km (395 miles)
Service ceiling: 9725m (31,900ft)
Dimensions: span 11.22m (36ft 10in); length 9.11m (29ft 11in); height 2.69m (8ft 10in)
Weight: 2624kg (5784lb) loaded
Armament: 8 x 7.7mm (0.303in) Browning MGs

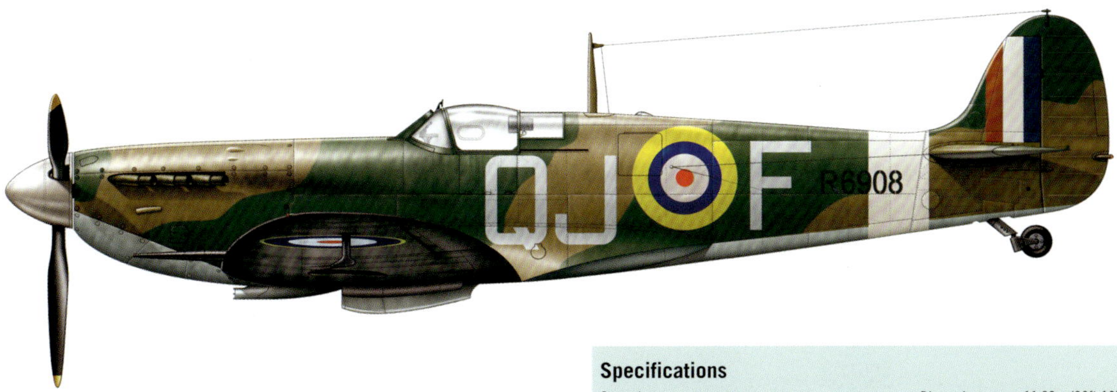

▲ Supermarine Spitfire Mk IB

No. 92 Squadron / RAF Fighter Command, 1940

The Spitfire Mk IB marked an important evolutionary step in the firepower of the Spitfire family, four of the the eight 7.7mm (0.303in) machine guns of the Spitfire Mk IA being replaced by two 20mm (0.79in) cannon.

Specifications

Crew: 1
Powerplant: 768kW (1030hp) Rolls-Royce Merlin III liquid-cooled V-12
Maximum speed: 582km/h (362mph)
Range: 636km (395 miles)
Service ceiling: 9725m (31,900ft)
Dimensions: span 11.22m (36ft 10in); length 9.11m (29ft 11in); height 2.69m (8ft 10in)
Weight: about 2624kg (5784lb) loaded
Armament: 2 x 20mm (0.79in) Hispano cannon and 4 x 7.7mm (0.303in) Browning MGs

RAF Component of the BEF

1939–40

Controlled by the so-called British Air Forces in France (BAFF), which also commanded the bomber squadrons of the Advanced Air Striking Force (AASF), the RAF Component of the British Expeditionary Force (BEF) provided reconnaissance and fighter capabilities.

THE RAF COMPONENT of the British Air Forces in France began life as Air Vice Marshal P. H. L. Playfair's Advanced Air Striking Force (AASF), which was formed on 24 August 1939 out of 10 light and medium bomber squadrons (Nos 12, 15, 40, 88, 103, 105, 142, 150, 218 and 226) of the Royal Air Force's No. 1 Group. This was transferred to France so that its warplanes would be able to operate against targets in western Germany.

The AASF was initially a command independent of the BEF, and as such reported directly to the Air Ministry in London. This arrangement soon proved to impractical, however, so on 15 January 1940 the AASF was made immediately subordinate to the headquarters of Air Marshal Sir Arthur S. 'Ugly' Barratt's BAFF. The BAFF also took under command Air Vice Marshal C. H. B. Blount's RAF Component of the BEF, whose Nos 1, 2, 4, 13, 16, 26, 53, 59, 73, 85, 87, 613 and 614 Squadrons comprised five Westland Lysander tactical reconnaissance, four Bristol Blenheim long-range reconnaissance and four (later six) Hawker Hurricane single-seat fighter units, the last later increased to six to protect installations and escort reconnaissances.

ORGANIZATION

British Air Forces in France
May 1940 (Air Marshal Arthur Barratt)

AASF (Air Vice Marshal P. H. L. Playfair)	Air Component (Air Vice Marshal C. H. B. Blount)	Squadrons detached

The BAFF's elements entered full-scale action, after limited skirmishing in the 'phoney war', on 10 May 1940 as the Germans launched their strategic offensive to the west through the Netherlands, Belgium and France. Like those of the French, the British air units suffered a stream of reverses and even disasters, for the aircraft types and numbers were wholly inadequate against a technically and tactically

Hawker Hurricane Mk IA

No. 1 Squadron / RAF Fighter Command, 1940

With its sturdy airframe and wide-track main landing gear units, the Hurricane was well suited for operations from the grass strips typical of Fighter Command's satellite airfields.

Specifications

Crew: 1	Service ceiling: 10,180m (33,400ft)
Powerplant: 768kW (1030hp) Rolls-Royce Merlin II liquid-cooled V-12	Dimensions: span 12.19m (40ft 0in); length 9.55m (31ft 4in); height 4.07m (13ft 4.5in)
Maximum speed: 496km/h (308mph)	Weight: 2820kg (6218lb) loaded
Range: 845km (525 miles)	Armament: 8 x 7.7mm (0.303in) Browning MGs

superior air adversary, and ground forces possessing excellent light anti-aircraft artillery. The Fairey Battle light bomber failed badly, most being shot down in the course of their first few missions, and the Blenheim light bomber did little better: during an attack on the bridges at Maastricht late in May, the AASF lost 40 out of 71 aircraft. As the Germans advanced, the BAFF retreated south, but by mid-June it was clear that defeat was imminent and the remnants of the BAFF began to withdraw to the UK, where the headquarters were eventually disbanded on 26 June 1940.

FIGHTER SQUADRONS OF THE RAF COMPONENT (SPRING 1940)

Unit	Type	Strength	Base
No. 1 Squadron	Hurricane I	12	Vassincourt
No. 73 Squadron	Hurricane I	12	n/a
No. 85 Squadron	Hurricane I	12	Lille
No. 87 Squadron	Hurricane I	12	n/a
No. 607 Squadron	Hurricane I	12	Merville
No. 615 Squadron	Hurricane I	12	Poix

1940 RAF Squadron

In 1940, before it had come into full and extended contact with the more advanced fighter tactics used by the *Luftwaffe*, the Royal Air Force had its 12-aircraft fighter squadrons organized in four flights each of three aircraft, and the flights and squadrons were trained to make formation attacks. Thereafter, the RAF quickly changed to a squadron organization and tactics most akin to those of the Germans, with the emphasis on pairs of fighters (leader and wingman) operating together.

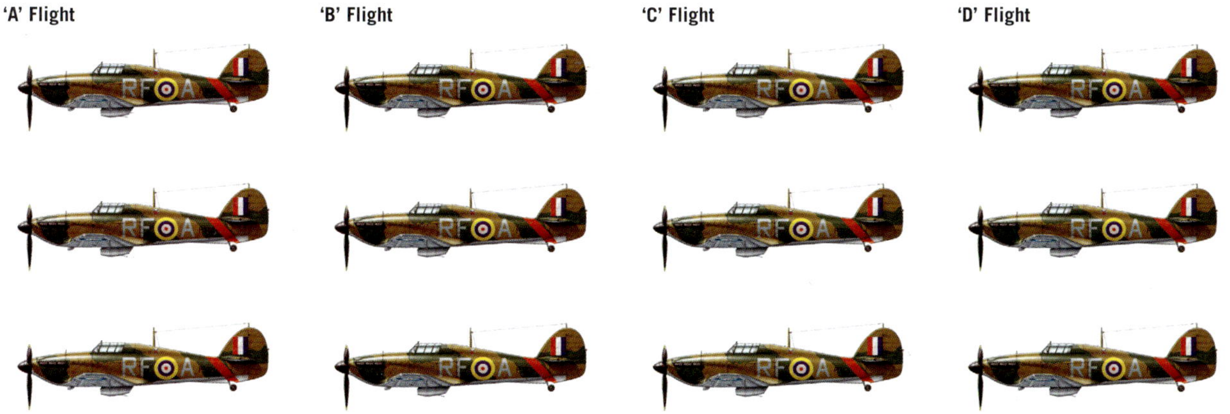

Battle of France

May 1940

From the first stages of the German offensive, the British air forces based in France had a torrid time of it. The reconnaissance and bomber units were savaged by technically more advanced adversaries, but the Hurricane fighter units were more effective and inflicted heavy losses.

The BAFF's airfields in northern France escaped essentially unscathed in the concentrated *Luftwaffe* attacks that accompanied the start of the invasion of the Low Countries and France on 10 May 1940. However, it was not long before the aircraft of the squadrons based on these airfields began to suffer a spate of losses as they came up against skilled German pilots. The enemy pilots had honed their capabilities in Spain and Poland, and were flying advanced warplanes, including (as the standard fighter) the first-class Messerschmitt Bf 109E single-engined warplane.

Although the Hawker Hurricane was a worthy, sturdy and easily maintained fighter, it was not in the same league as the Bf 109. The BAFF's other primary fighter, although in the last stages of being phased out of service, was the Gloster Gladiator biplane with only four rather than eight machine guns, and fixed landing gear.

As always, the fighters were tasked with the protection of key installations in France against German bomber attack, and the escort of Bristol Blenheim twin-engined light bombers and long-range reconnaissance aircraft, plus Fairey Battle single-engined light bombers and Westland Lysander battlefield reconnaissance aircraft.

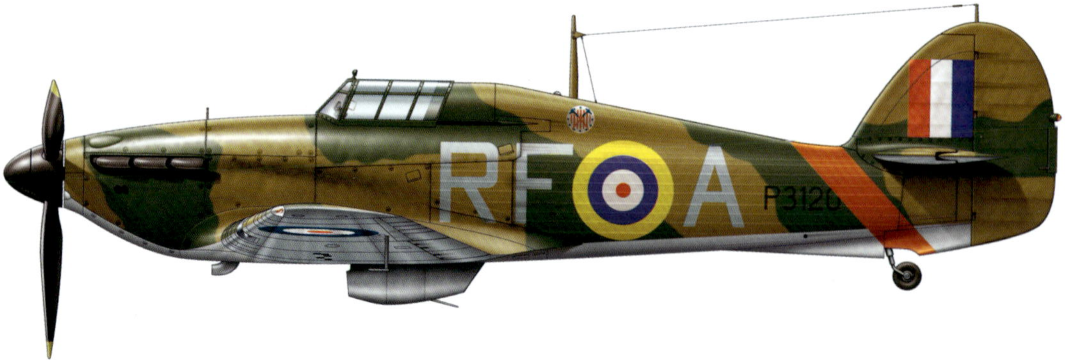

▲ Hawker Hurricane Mk I

No. 303 Squadron / RAF Fighter Command, Northolt, August 1940

The Hawker Hurricane was an excellent blend of old (steel tube in the structure and largely fabric covering) and new (low-set cantilever wing, enclosed cockpit, retractable main landing gear units and eight-gun main armament). In the Battle of Britain, the Hurricane was Fighter Command's highest-scoring fighter.

Specifications

Crew: 1
Powerplant: 768kW (1030hp) Rolls-Royce Merlin II liquid-cooled V-12
Maximum speed: 496km/h (308mph)
Range: 845km (525 miles)
Service ceiling: 10,180m (33,400ft)
Dimensions: span 12.19m (40ft 0in); length 9.55m (31ft 4in); height 4.07m (13ft 4.5in)
Weight: 2820kg (6218lb) loaded
Armament: 8 x 7.7mm (0.303in) Browning MGs

Specifications

Crew: 1
Powerplant: 768kW (1030hp) Rolls-Royce Merlin II liquid-cooled V-12
Maximum speed: 496km/h (308mph)
Range: 845km (525 miles)
Service ceiling: 10,180m (33,400ft)
Dimensions: span 12.19m (40ft 0in); length 9.55m (31ft 4in); height 4.07m (13ft 4.5in)
Weight: 2820kg (6218lb) loaded
Armament: 8 x 7.7mm (0.303in) Browning MGs

▲ Hawker Hurricane Mk IA

No. 87 Squadron / No. 60 Wing

With the introduction of Hurricane Mk II variants armed with cannon, aircraft with machine-gun armament received the letter 'A' as a suffix to their designation. The Hurricane was very sturdy, and later became an excellent fighter-bomber.

With escort, the Blenheim could achieve limited successes, but the Battle and Lysander were very quickly revealed as fatally vulnerable not only to fighter attack but also to the capabilities of the anti-aircraft guns that the Germans deployed very strongly in calibres between 20mm (0.79in) and 88mm (3.465in).

Air support role

While fighters despatched by the UK-based Fighter Command attempted to provide air support for the Dutch and Belgians, the BAFF's 416 aircraft, including 96 Hurricanes in six squadrons, were facing a losing battle slightly further to the west, even though they were reinforced by another three Hurricane units, Nos 3, 79 and 501 Squadrons.

The nature of the British air element's task is indicated by the fact that on the day of its arrival, on 10 May, No. 501 Squadron refuelled and immediately rose into combat, two of the Hurricanes becoming embroiled with 40 or more Heinkel He 111 bombers. During the day, the BAFF flew 161 fighter sorties, 81 leading to combat and resulting in claims for 36 German aircraft downed at the cost of two Hurricanes lost.

As the campaign progressed, the kill/loss ratio veered steadily in favour of the Germans as the Allies' loss of territory compounded servicing and maintenance problems, and as pilots became increasingly exhausted. Rightly believing that France would be defeated, Air Chief Marshal Sir Hugh Dowding, commanding Fighter Command, refused to allow Supermarine Spitfire squadrons to be transferred to France, and the BAFF was pulled out late in June so that the RAF could prepare for the Battle of Britain. The Battle of France had cost the RAF 959 aircraft, including 477 invaluable fighters.

▲ BAFF Gladiators

Pilots from No. 615 Squadron sit besides their Gladiator Mk II fighters as part of the Air Component of the British Expeditionary Force, early in 1940. The Gladiator proved no match for the more modern warplanes of the *Luftwaffe*.

▲ Bristol Blenheim Mk IF

No. 25 Squadron / RAF Fighter Command, Hawkinge, Kent, summer 1940

The Blenheim Mk IF was an interim fighter created out of the Blenheim Mk I light bomber by the addition of a ventral pack carrying four 7.7mm (0.303in) Browning fixed forward-firing machine guns.

Specifications

Crew: 2

Powerplant: 2 x 626kW (840hp) Bristol Mercury VIII air-cooled 9-cylinder radial

Maximum speed: 447km/h (278mph)

Range: 1690km (1050 miles)

Service ceiling: 8315m (27,280ft)

Dimensions: span 17.17m (56ft 4in); length 12.12m (39ft 9in); height 3.00m (9ft 10)

Weight: 5534kg (12,200lb) loaded

Armament: 5 x 7.7mm (0.303in) Browning MGs and 1 x 7.7mm (0.303in) Vickers 'K' MG

Battle of Britain

July–October 1940

The Battle of Britain was a major turning point in World War II, for it was the first occasion on which the Germans suffered a strategic defeat. To launch a seaborne invasion of the UK, the Germans needed air supremacy, but failed to win this in the Battle of Britain.

THE BATTLE OF BRITAIN was the first major battle to be fought entirely by air forces. The Germans thought that a successful amphibious assault on the UK could not be made until the RAF had been neutralized, and this led to the battle dated by British historians to the period between 10 July and 31 October 1940, in which the Germans' major daylight effort was made. German historians usually record the battle as starting in mid-August 1940 and ending in May 1941, when the German bomber arm was withdrawn for the invasion of the USSR.

In an effort to finish the war in the west, on 16 July Hitler ordered the rapid preparation of the *Seelöwe* (Sea Lion) invasion for launch in mid-August. The *Luftwaffe* was now facing an opponent more capable than the French Air Force, for the RAF was well-trained, well coordinated, high in morale and equipped with fighters fully the equal of the Messerschmitt Bf 109E single-engined and Bf 110 twin-engined fighters.

The primary weight of the RAF's fighting fell on the Hurricane Mk I, but the Germans now came to appreciate that the newer Spitfire Mk I was a superb fighter. The British had fewer experienced pilots at

ORGANIZATION

RAF Fighter Command
July 1940

No.11 Group (Air Vice Marshal Keith Park)	No. 12 Group (Air Vice Marshal Trafford Leigh-Mallory)	No. 13 Group (Air Vice Marshal Richard Saul)

NO.12 GROUP: SINGLE-ENGINED FIGHTER SQUADRONS (JULY 1940)

Unit	Type	Strength	Serviceable
No.19 Squadron	Spitfire	8	5
No. 264 Squadron	Defiant	11	7
No. 66 Squadron	Spitfire	12	4
No. 242 Squadron	Hurricane	10	4
No. 222 Squadron	Spitfire	12	4
No. 46 Squadron	Hurricane	15	3
No. 611 Squadron	Spitfire	3	11

Hurricane fighters

A pair of Mk I Hurricanes from No. 111 Squadron prepare to do battle with the *Luftwaffe* over southeast England, summer 1940.

the start of the battle, and it was this shortage of trained pilots which became the greatest concern for Air Chief Marshal Sir Hugh Dowding. Drawing from regular RAF forces as well as the Auxiliary Air Force and the Volunteer Reserve, the British could muster some 1103 fighter pilots on 1 July. The Germans could muster 1450 fighter pilots, and these were generally more experienced.

In the early phases of the battle, the RAF was hindered by its reliance on dated formations. These restricted squadrons to tight 12 aircraft formations composed of three-aircraft 'sections' in tight 'vics'.

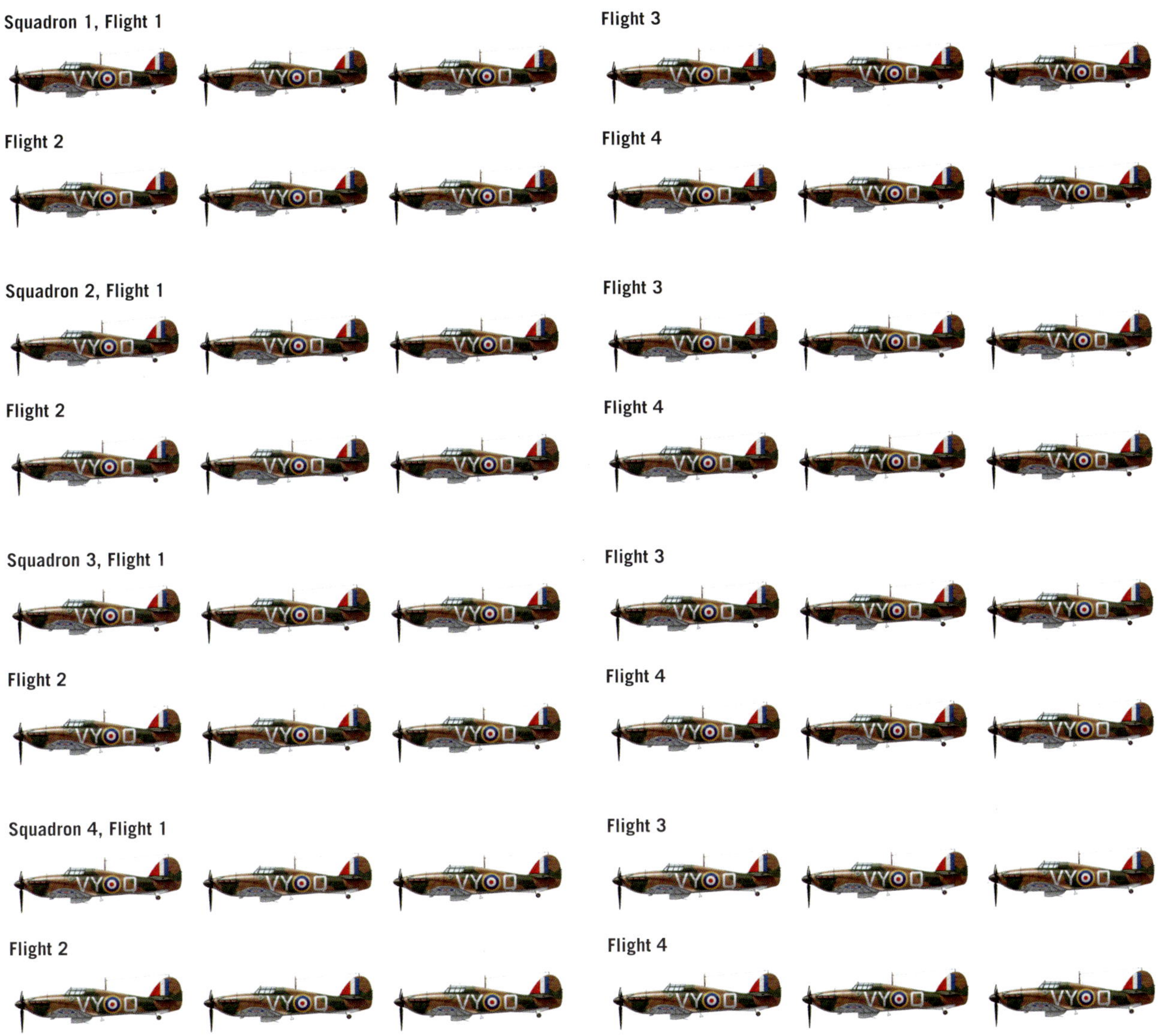

▲ 1940 RAF 'Big Wing'

The brainchild of Squadron Leader D. R. S. Bader and Air Vice Marshal Leigh-Mallory of No. 12 Group, the 'big wing' was based on the idea that four squadrons operating together would inflict more damage than four squadrons operating independently. However, it took so long to assemble the 'big wing' in the air that it could not intercept the Germans bombers before they had attacked London, and Air Vice Marshal Park of No. 11 Group, supported by Dowding, argued that any type of pre-attack interception was better in degrading the bombers' effectiveness.

Specifications

Crew: 1
Powerplant: 768kW (1030hp) Rolls-Royce Merlin III liquid-cooled V-12
Maximum speed: 582km/h (362mph)
Range: 636km (395 miles)
Service ceiling: 9725m (31,900ft)
Dimensions: span 11.22m (36ft 10in); length 9.11m (29ft 11in); height 2.69m (8ft 10in)
Weight: 2624kg (5784lb) loaded
Armament: 8 x 7.7mm (0.303in) Browning MGs

▲ **Supermarine Spitfire Mk IA**

No. 66 Squadron / RAF Fighter Command, 1940

With 12 serviceable and four unserviceable aircraft on 1 July 1940, No. 66 Squadron was part of No. 12 Group and based at Coltishall in East Anglia.

Specifications

Crew: 1
Powerplant: 768kW (1030hp) Rolls-Royce Merlin III liquid-cooled V-12
Maximum speed: 582km/h (362mph)
Range: 636km (395 miles)
Service ceiling: 9725m (31,900ft)
Dimensions: span 11.22m (36ft 10in); length 9.11m (29ft 11in); height 2.69m (8ft 10in)
Weight: 2624kg (5784lb) loaded
Armament: 8 x 7.7mm (0.303in) Browning MGs

▲ **Supermarine Spitfire Mk IA**

No. 602 Squadron / RAF Fighter Command, 1940

With 12 serviceable and four unserviceable aircraft on 1 July 1940, No. 602 Squadron was part of No. 13 Group and based at Drem in Scotland.

With four sections flying together in tight formation, only the squadron leader at the front was free to search for the enemy. British training also emphasized formulaic attacks by sections breaking away in sequence. Fighter Command recognized the weaknesses of this rigid structure early in the battle, but it was felt too risky to change tactics as the fighting continued. A compromise allowed the squadron formations to fly looser formations with one or two 'weavers' flying above and behind for greater observation and rear protection. After the battle, RAF pilots adopted a variant on the German *Schwarm* (two pairs, each consisting of a leader and a wingman). Each *Schwarm* in a *Staffel* (squadron) flew staggered and with plenty of room between them, making the formation difficult to spot at longer ranges and allowing for greater flexibility.

The *Luftwaffe* regrouped after the Battle of France into three *Luftflotten* (air fleets) on the UK's southern and northern flanks. *Luftflotte 2,* commanded by *Generalfeldmarschall* Albert Kesselring, was responsible for the bombing of south-east England and the London area. *Luftflotte 3,* under *Generalfeldmarschall* Hugo Sperrle, targeted the West Country, Midlands and north-west England. *Luftflotte 5,* under *Generaloberst* Hans-Jürgen Stumpff from his headquarters in Norway, targeted the north of England and Scotland.

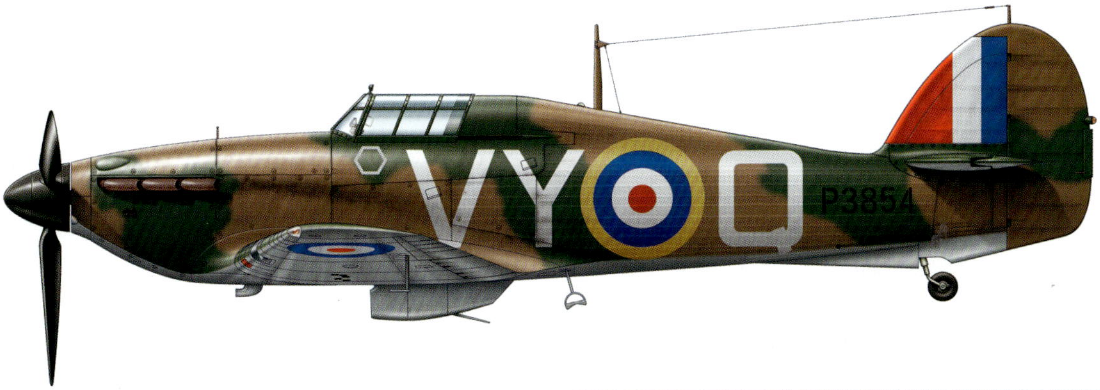

▲ **Hawker Hurricane Mk IA**

No. 85 Squadron / RAF Fighter Command, August 1940

With 15 serviceable and three unserviceable aircraft on 1 July 1940, No. 85 Squadron was part of No. 12 Group and based at Martlesham in East Anglia.

Specifications

Crew: 1
Powerplant: 768kW (1030hp) Rolls-Royce Merlin II liquid-cooled V-12
Maximum speed: 496km/h (308mph)
Range: 845km (525 miles)
Service ceiling: 10,180m (33,400ft)
Dimensions: span 12.19m (40ft 0in); length 9.55m (31ft 4in); height 4.07m (13ft 4.5in)
Weight: 2820kg (6218lb) loaded
Armament: 8 x 7.7mm (0.303in) Browning MGs

No. 10 Group

June–December 1940

A late arrival in the establishment of RAF Fighter Command, No. 10 Group helped to lift the operational burden from the all-important No. 11 to its east, and also provided this latter group with replacement pilots and aircraft to keep up the fighter strength in south-east England.

THE FIRST No. 10 Group of the Royal Air Force was formed on 1 April 1918 in No. 2 Area. On 8 May of the following year, it was transferred to the South-Western Area, and in 1918 it was further transferred to the Coastal Area, with which the group remained until it was disbanded on 18 January 1932.

ORGANIZATION

No. 10 Group, RAF Fighter Command
September 1940

- Pembrey Sector (one squadron)
- Filton Sector (three squadrons)
- St Eval Sector (two squadrons)
- Middle Wallop Sector (six squadrons)

New group

No. 10 Group was re-established on 1 June 1940 within Air Chief Marshal Sir Hugh Dowding's Fighter Command. The rationale for this expansion of Fighter Command's subordinate formations was to allow what was now the neighbouring No. 11 Group, previously responsible for the protection of the whole of southern England, to concentrate its efforts on just south-east England, which would clearly be the area most centrally threatened by German air attack from bases in occupied northern France, Belgium and the Netherlands. Thus the new No. 10 Group was allocated south-west England and the portions of England and Wales flanking the Bristol Channel as its area of responsibility.

Under the command of Air Vice Marshal Sir Christopher Brand, No. 10 Group supported No. 11 Group in the Battle of Britain by rotating squadrons, transferring extra fighter support when this was required, and making more pilots available when necessary. It is notable that Air Vice Marshal Keith Park, commanding the decisive No. 11 Group,

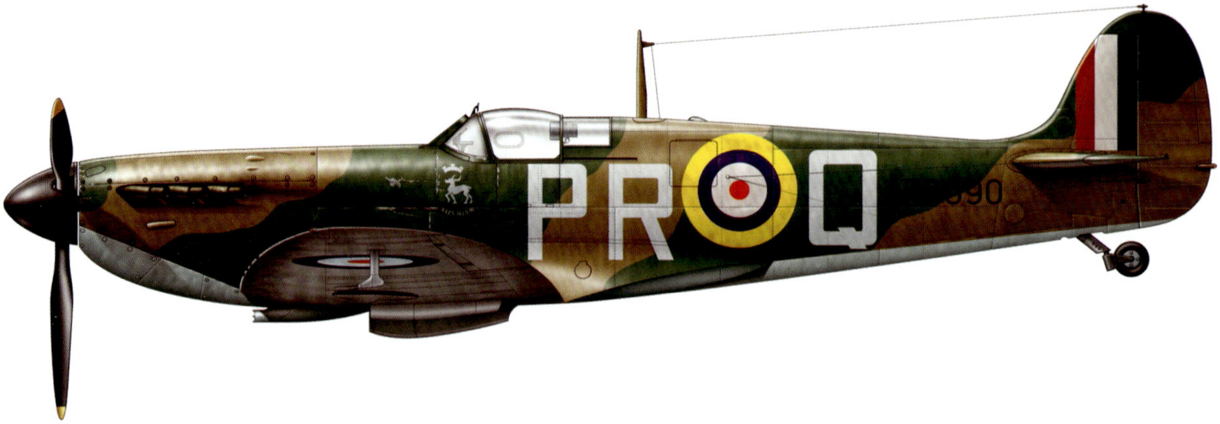

▲ **Supermarine Spitfire Mk IA**

No. 609 Squadron / 10 Group

No. 609 Squadron was formed at RAF Yeadon, now Leeds Bradford International Airport, on 10 February 1936, as one of the 20 flying Squadrons of the Royal Auxiliary Air Force. During the Battle of Britain, the squadron moved to RAF Middle Wallop, west of London, as part of Fighter Command's efforts to defend the south coast of England.

Specifications

Crew: 1
Powerplant: 768kW (1030hp) Rolls-Royce Merlin III liquid-cooled V-12
Maximum speed: 582km/h (362mph)
Range: 636km (395 miles)
Service ceiling: 9725m (31,900ft)
Dimensions: span 11.22m (36ft 10in); length 9.11m (29ft 11in); height 2.69m (8ft 10in)
Weight: 2624kg (5784lb) loaded
Armament: 8 x 7.7mm (0.303in) Browning MGs

enjoyed a far better relationship with Brand than with Air Vice Marshal Trafford Leigh-Mallory of No. 12 Group.

As well as providing support for No. 11 Group, No. 10 Group had a number of squadrons equipped with aircraft whose obsolescence rendered them unsuitable for active use in the combat arena of the

NO. 10 GROUP SQUADRONS, 1 SEPTEMBER 1940

Unit	Type	Strength	Serviceable
No. 87 Squadron	Hurricane	15	9
No. 92 Squadron	Spitfire	16	12
No. 152 Squadron	Spitfire	16	12
No. 213 Squadron	Hurricane	15	8
No. 234 Squadron	Spitfire	17	12
No. 236 Squadron	Blenheim	17	12
No. 238 Squadron	Hurricane	15	11
No. 249 Squadron	Hurricane	16	15
No. 604 Squadron	Blenheim	14	11
No. 609 Squadron	Spitfire	14	11

▼ **Bristol Beaufighter**

The Beaufighter Mk IF was the first effective night-fighter in British service, a role for which it was cleared in July 1940. Large and powerful, the Beaufighter possessed adequate performance, but carried AI. Mk IV radar and the fixed forward-firing armament of four 20mm (0.79in) cannon. This was a machine of No. 604 Squadron.

Battle of Britain. These types included the Gloster Gladiator biplane and the Boulton Paul Defiant turret fighter.

No. 10 Group layout

No. 10 Group had its headquarters at Box in Wiltshire, and its primary sectors in September were Pembrey, Filton (with other airfields at Exeter, Bibury and Colerne), St Eval (with another airfield at Roborough), and Middle Wallop (with other airfields at Warmwell and Boscombe Down). There were also operational training units at Aston Down, Sutton Bridge and Hawarden airfields. After the end of the Battle of Britain, No. 10 Group was responsible for offensive missions into north-west France by the Westland Whirlwind twin-engined attack fighter, and also provided fighter cover for the all-important convoys of merchant ships approaching and leaving the British Isles via the South-Western Approaches. Any pilots posted to a squadron of No. 10 Group from either No. 12 Group or No. 13 Group knew that they would soon be transferred to No. 11 Group, where the weight of the growing British air offensive against German fighters and their bases in northern France was concentrated from early in 1941. These pilots, therefore, took every advantage of No. 10 Group's comparatively safe area of operation to hone their skills in preparation for a more active level of service.

It is worth noting that No. 10 Group was finally reabsorbed into No. 11 Group on 2 May 1945, only days before the end of World War II in Europe.

No. 11 Group

July–September 1940

The story of No. 11 Group is in large measure the story of Fighter Command in the Battle of Britain. Based in the south-east of England, the group bore the brunt of the German air offensive during the fighter and bomber phases of the battle, and emerged scarred but unbeaten.

A No. 11 Group was first formed on 1 April 1918 in No. 2 Area, and was transferred to the South-Western Area the next month on 8 May.

The group was disbanded on 17 May of the same year, only to be re-formed in the North-Western Area on 22 August. In May 1920, No. 11 Group was reduced to No. 11 Wing. Then on 14 July 1936, No. 11 Group became the first group to be established within Fighter Command, and had responsibility for the defence of southern England, including London.

Group organization

The group was organized using the Dowding system of fighter control. The group's HQ was at RAF Uxbridge, from where commands were passed to the various sector airfields, each of which coordinated several airfields and fighter squadrons. No. 11 Group's sectors (A, B, C, D, E, F, Y and Z) airfields were Tangmere, Kenley, Biggin Hill, Hornchurch, North Weald, Debden, Middle Wallop and Northolt.

New commander

Air Vice Marshal Keith Park assumed command of No. 11 Group on 20 April 1940 and led it throughout its period of enduring fame in the Battle of Britain. Pilots posted to No. 11 Group's squadrons

ORGANIZATION

No. 11 Group (Air Vice Marshal Keith Park)
July 1940

- Biggin Hill Sector
- North Weald Sector
- Kenley Sector
- Northolt Sector
- Hornchurch Sector
- Tangmere Sector
- Filton Sector
- Middle Wallop Sector

knew they would see extensive action, while the pilots leaving the group knew that they were going to comparatively safer duty.

Enjoying the full support of the commanders of Nos 10 and 13 Groups, Park received only modest support from No. 12 Group's Air Vice Marshal Trafford Leigh-Mallory, who did not agree with Park's tactics.

Restructured command

Park orchestrated his command very carefully in the five phases of the battle. In the first (10 July–7 August), the British and Germans lost 169 and 192 aircraft during fighting over the English Channel and the British ports along it.

In the second phase (8–23 August), the losses were 303 and 403 respectively as the Germans attacked radar stations and forward fighter bases.

In the third (24 August–6 September) the losses were 262 and 378 as the Germans attacked inland fighter bases and fighter factories to force the RAF into the air.

In the fourth (7–30 September), the losses were 380 and 435 in German daylight attacks on London, before the Germans switched to fighter-bomber raids in the fifth phase (1–31 October), when they lost 325 aircraft to the British 265, and thereafter

NO. 11 GROUP FIGHTER SQUADRONS (1 AUGUST 1940)

Unit	Type	Strength	Serviceable
No. 1 Squadron	Hurricane	16	13
No. 17 Squadron	Hurricane	19	14
No. 25 Squadron	Blenheim	14	7
No. 32 Squadron	Hurricane	15	11
No. 41 Squadron	Spitfire	16	10
No. 43 Squadron	Hurricane	19	18
No. 56 Squadron	Hurricane	17	5
No. 64 Squadron	Spitfire	16	12
No. 65 Squadron	Spitfire	16	11
No. 74 Squadron	Spitfire	15	12
No. 85 Squadron	Hurricane	18	12
No. 111 Squadron	Hurricane	12	10
No. 145 Squadron	Hurricane	17	10
No. 151 Squadron	Hurricane	18	13
No. 257 Squadron	Hurricane	15	10
No. 266 Squadron	Spitfire	18	13
No. 501 Squadron	Hurricane	16	11
No. 600 Squadron	Blenheim	15	9
No. 601 Squadron	Hurricane	18	14
No. 610 Squadron	Spitfire	15	12
No. 615 Squadron	Hurricane	16	14

Specifications

Crew: 1
Powerplant: 768kW (1030hp) Rolls-Royce Merlin III liquid-cooled V-12
Maximum speed: 582km/h (362mph)
Range: 636km (395 miles)
Service ceiling: 9725m (31,900ft)
Dimensions: span 11.22m (36ft 10in); length 9.11m (29ft 11in); height 2.69m (8ft 10in)
Weight: 2624kg (5784lb) loaded
Armament: 8 x 7.7mm (0.303in) Browning MGs

▲ **Supermarine Spitfire Mk IA**

No. 74 Squadron / RAF Fighter Command, 1940

In July 1940, No. 74 Squadron was based at Hornchurch in No. 11 Group.

switched to night bombing. With the battle over, Leigh-Mallory and Air Marshal Sholto Douglas managed to have Park and Air Chief Marshal Sir Hugh Dowding removed, moving into their positions at the head of No. 11 Group and Fighter Command respectively.

Specifications

Crew: 2
Powerplant: 2 x 1119kW (1500hp) Bristol Hercules XI air-cooled engine
Maximum speed: 492km/h (306mph)
Range: 2414km (1500 miles)
Service ceiling: 8810m (28,900ft)
Dimensions: span 17.63m (57ft 10in); length 12.60m (41ft 4in); height 4.82m (15 ft 10in)
Weight: 9435kg (21,100lb) loaded
Armament: 4 x 20mm (0.79in) cannon and 6 x 7.7mm (0.303in) MGs

▲ Bristol Beaufighter Mk IF

No. 25 Squadron / RAF Fighter Command

The Beaufighter multi-role warplane saw its first service as a heavy fighter, and most especially as a night-fighter. As such, it was the RAF's first effective means of tackling German night-bombers.

▲ Hawker Hurricane Mk I

No. 249 Squadron / No. 11 Group

Based at RAF North Weald, this Hurricane was flown by Squadron Leader John Grandy. The squadron was formed as a fighter unit on 16 May 1940 at Church Fenton. It was equipped with Spitfires for the first month, but these were changed to Hurricanes in June.

Specifications

Crew: 1
Powerplant: 768kW (1030hp) Rolls-Royce Merlin II liquid-cooled V-12
Maximum speed: 496km/h (308mph)
Range: 845km (525 miles)
Service ceiling: 10,180m (33,400ft)
Dimensions: span 12.19m (40ft 0in); length 9.55m (31ft 4in); height 4.07m (13ft 4.5in)
Weight: 2820kg (6218lb) loaded
Armament: 8 x 7.7mm (0.303in) Browning MGs

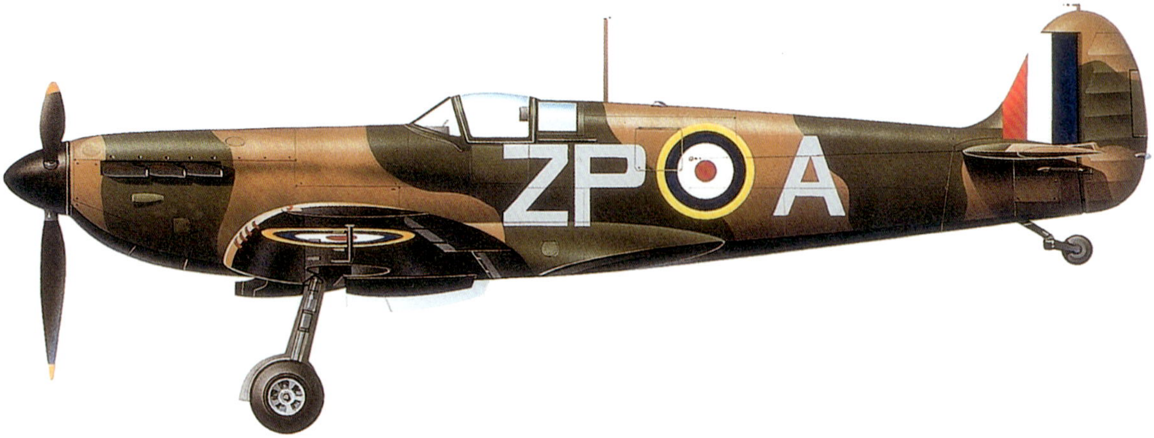

▲ Supermarine Spitfire Mk IA

No. 74 Squadron / Fighter Command, 1940

Led by Squadron Leader F. L. White, No. 74 Squadron was based at Hornchurch as one of the three Spitfire units operating in No. 11 Group's Hornchurch sector. The squadron had 15 aircraft at this time, 12 of them serviceable.

Specifications

Crew: 1
Powerplant: 768kW (1030hp) Rolls-Royce Merlin III liquid-cooled V-12
Maximum speed: 582km/h (362mph)
Range: 636km (395 miles)
Service ceiling: 9725m (31,900ft)
Dimensions: span 11.22m (36ft 10in); length 9.11m (29ft 11in); height 2.69m (8ft 10in)
Weight: 2624kg (5784lb) loaded
Armament: 8 x 7.7mm (0.303in) Browning MGs

▲ Hawker Hurricane Mk IA

No. 501 Squadron / 11 Group / RAF Fighter Command, 1940

Based at Gravesend in the Biggin Hill Sector of No. 11 Group on 1 September 1940, No. 501 Squadron was led by Squadron Leader H. A. V. Hogan and on this day had a strength of 17 Hurricane fighters, only 11 of them serviceable.

Specifications

Crew: 1
Powerplant: 768kW (1030hp) Rolls-Royce Merlin II liquid-cooled V-12
Maximum speed: 496km/h (308mph)
Range: 845km (525 miles)
Service ceiling: 10,180m (33,400ft)
Dimensions: span 12.19m (40ft 0in); length 9.55m (31ft 4in); height 4.07m (13ft 4.5in)
Weight: 2820kg (6218lb) loaded
Armament: 8 x 7.7mm (0.303in) Browning MGs

1940 RAF Spitfire ‘Big Wing’

The rationale of the ‘big wing’, according to its protagonists, was that while its assembly time might deny it the opportunity to effect an interception before the bombers attacked, it would deliver so heavy a blow as the bombers streamed home that the Germans’ ability to sustain the bomber offensive would soon be degraded. This factor would, in turn, ease the task of the defending fighters and lead to a steadily escalating German loss rate.

Squadron 1, ‘A’ Flight

‘C’ Flight

‘B’ Flight

‘D’ Flight

Squadron 2, ‘A’ Flight

‘C’ Flight

‘B’ Flight

’D’ Flight

Squadron 3, ‘A’ Flight

‘C’ Flight

‘C’ Flight

‘D’ Flight

Squadron 4, ‘A’ Flight

‘C’ Flight

‘B’ Flight

‘D’ Flight

Squadron 5, ‘A’ Flight

‘C’ Flight

‘B’ Flight

‘D’ Flight

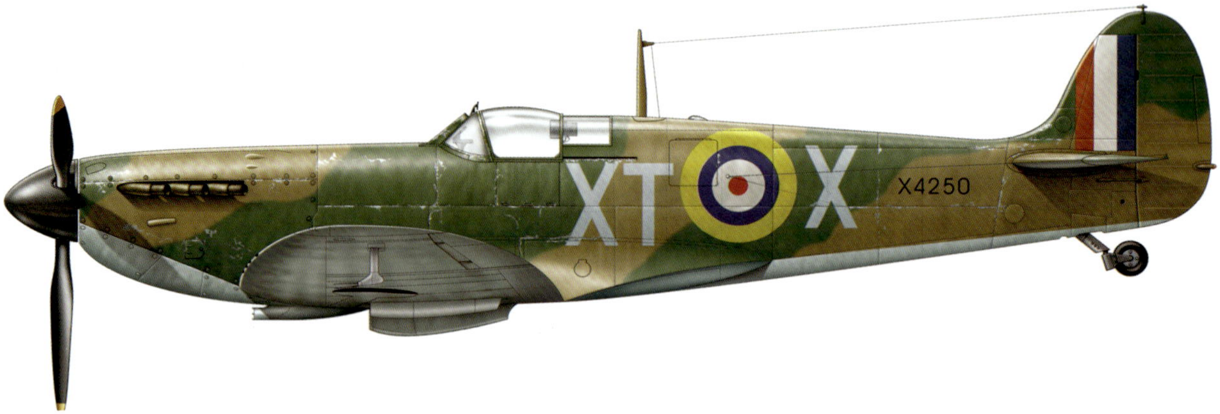

▲ **Supermarine Spitfire Mk IA**

No. 603 Squadron / RAF Fighter Command, 1940

In 1 July 1940, No. 603 Squadron was part of No. 13 Group, and based at Turnhouse in Scotland with flights detached to Dyce and Montrose.

Specifications

Crew: 1
Powerplant: 768kW (1030hp) Rolls-Royce Merlin III liquid-cooled V-12
Maximum speed: 582km/h (362mph)
Range: 636km (395 miles)
Service ceiling: 9725m (31,900ft)
Dimensions: span 11.22m (36ft 10in); length 9.11m (29ft 11in); height 2.69m (8ft 10in)
Weight: 2624kg (5784lb) loaded
Armament: 8 x 7.7mm (0.303in) Browning MGs

▲ **Supermarine Spitfire Mk I**

No. 66 Squadron / 11 Group

Based at RAF Biggin Hill, the Spitfires of 66 Squadron were commanded by Squadron Leader Rupert Leigh. The squadron received its first Spitfires in late 1938, and flew operational patrols over Dunkirk in the withdrawal from France in May 1940. The squadron joined 11 Group in September 1940.

Specifications

Crew: 1
Powerplant: 768kW (1030hp) Rolls-Royce Merlin III liquid-cooled V-12
Maximum speed: 582km/h (362mph)
Range: 636km (395 miles)
Service ceiling: 9725m (31,900ft)
Dimensions: span 11.22m (36ft 10in); length 9.11m (29ft 11in); height 2.69m (8ft 10in)
Weight: 2624kg (5784lb) loaded
Armament: 8 x 7.7mm (0.303in) Browning MGs

▲ **On patrol**

Spitfire Mk IA fighters of No. 11 Group's No. 610 Squadron, based at RAF Biggin Hill, in June 1940. The aircraft are flying in flights of three aircraft.

No. 12 Group

JULY–SEPTEMBER 1940

Covering the Midlands of England, No. 12 Group was not as hard pressed as No. 11 Group, and was thus well placed to come to the aid of its southern neighbour, which had more squadrons but always, despite the advantages offered by radar, lacked time to assemble in numbers.

NO. 12 GROUP of the Royal Air Force first came into existence during April 1918 at Cranwell, Lincolnshire, as part of the No. 3 Area. On 8 May 1918, the group transferred to the Midland Area, and then on 18 October of the same year to the Northern Area. On 1 November 1918, the group became the RAF (Cadet) College.

Reconstituted group

No. 11 Group was reconstituted on 1 April 1937 as part of Fighter Command. In World War II, No. 12 Group was entrusted with the task of providing air defence for the Midlands of England, and as such was of importance second only to that of No. 11 Group. As No. 12 Group's area of responsibility included many industrial areas, it received a large number of German bombing attacks.

The commander of No. 12 group during the summer and autumn of 1940 was Air Vice Marshal Trafford Leigh-Mallory, an ambitious officer who had assumed command of the group in December 1937. Despite his length of service in the RAF, Leigh-Mallory had been passed over for command of the more important No. 11 Group in favour of Air Vice Marshal Keith Park. Feeling that he had been slighted, Leigh-Mallory thereafter had a distinctly cool relationship with Park, hardly the ideal situation for the commanders of the UK's two most important air defence groups.

'Big wing' tactics

In September 1940, No. 12 Group's sectors were Duxford, Coltishall, Wittering, Digby, Kirton-in-Lindsey (three airfields) and Church Fenton (three airfields) with elements of 17 squadrons. In addition to providing regional air defence over the Midlands, No. 12 Group was also allocated the task of providing cover for No. 11 Group's airfields during the Battle of Britain, but on several occasions these vital

NO. 12 GROUP FIGHTER SQUADRONS (1 SEPTEMBER 1940)

Unit	Type	Strength	Serviceable
No. 19 Squadron	Spitfire	15	11
No. 23 Squadron	Blenheim	17	11
No. 29 Squadron	Blenheim	14	10
No. 46 Squadron	Hurricane	17	15
No. 74 Squadron	Spitfire	16	11
No. 222 Squadron	Spitfire	16	10
No. 229 Squadron	Hurricane	16	12
No. 242 Squadron	Hurricane	15	11
No. 264 Squadron	Defiant	15	8
No. 266 Squadron	Spitfire	12	8
No. 310 Squadron	Hurricane	14	10
No. 611 Squadron	Spitfire	18	12

ORGANIZATION

No. 12 Group (Air Vice Marshal Trafford Leigh-Mallory)
August 1940

- Duxford Sector
- Coltishall Sector
- Kirton-in-Lindsey Sector
- Digby Sector
- Wittering Sector

Specifications

Crew: 2
Powerplant: 768kW (1030hp) Rolls-Royce Merlin III liquid-cooled V-12
Maximum speed: 489km/h (304mph)
Range: 748km (465 miles)
Service ceiling: 9250m (30,350ft)
Dimensions: span 11.99m (39ft 4in); length 10.77m (35ft 4in); height 3.71m (12ft 2in)
Weight: 3773kg (8318lb) loaded
Armament: 4 x 7.7mm (0.303in) Browning MGs in the power-operated dorsal turret

▲ Boulton Paul Defiant Mk I

No. 264 Squadron / RAF Fighter Command, mid-1940

This Defiant Mk I was flown by Squadron Leader P. A. Hunter. The Defiant scored some initial success when the nature of its turreted armament was not comprehended, but once the Germans knew that the aircraft lacked both fixed forward-firing guns and agility, they were able to shoot the type down with ease.

installations were left undefended. When Park complained about it, Leigh-Mallory responded that greater results would accrue from the implementation of the 'big wing' tactic, but that the assembly of a concentrated force in the air took time.

The 'big wing' was certainly an effective air fighting force whenever it was actually assembled and managed to effect an interception with the massed enemy air units, but this happened on only a limited number of occasions. Even so, opponents of Park and Air Chief Marshal Sir Hugh Dowding were able to exploit the low incidence of these successes after the end of the battle and so engineer the removal of Park and Dowding, who were replaced by Leigh-Mallory and Air Marshal Sir William Sholto Douglas respectively.

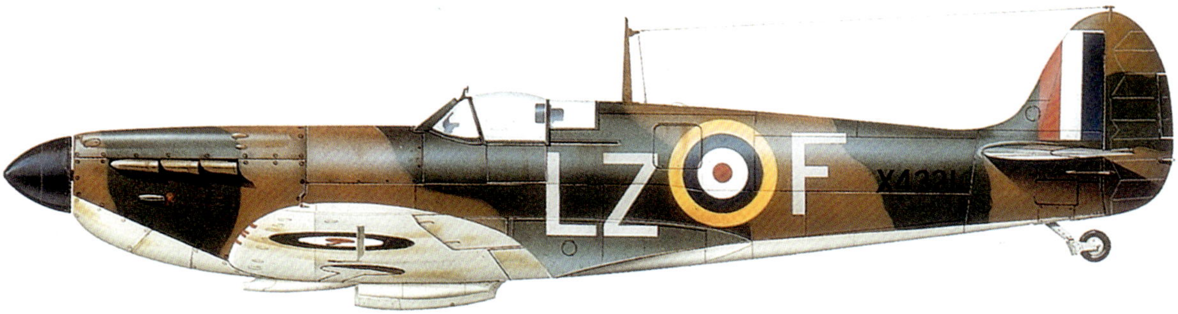

Specifications

Crew: 1
Powerplant: 768kW (1030hp) Rolls-Royce Merlin III liquid-cooled V-12
Maximum speed: 582km/h (362mph)
Range: 636km (395 miles)
Service ceiling: 9725m (31,900ft)
Dimensions: span 11.22m (36ft 10in); length 9.11m (29ft 11in); height 2.69m (8ft 10in)
Weight: 2624kg (5784lb) loaded
Armament: 8 x 7.7mm (0.303in) Browning MGs

▲ **Supermarine Spitfire Mk IA**

No. 66 Squadron / RAF Fighter Command, 1940

Led by Squadron Leader R. H. A. Leigh, No. 66 Squadron had 16 aircraft, 12 of them serviceable, on 1 August 1940. The squadron was part of No. 12 Group's strength, and was based at Coltishall as one of the two squadrons in the Coltishall sector. The other unit was No. 242 Squadron with Hurricanes.

Specifications

Crew: 1
Powerplant: 768kW (1030hp) Rolls-Royce Merlin III liquid-cooled V-12
Maximum speed: 582km/h (362mph)
Range: 636km (395 miles)
Service ceiling: 9725m (31,900ft)
Dimensions: span 11.22m (36ft 10in); length 9.11m (29ft 11in); height 2.69m (8ft 10in)
Weight: about 2624kg (5784lb) loaded
Armament: 2 x 20mm (0.79in) Hispano cannon and 4 x 7.7mm (0.303in) Browning MGs

▲ **Supermarine Spitfire Mk IB**

No. 19 Squadron / RAF Fighter Command, 1940

No. 19 Squadron was the first unit to receive Spitfire fighters, in August 1938. At the beginning of September 1940, the unit was based at Fowlmere in the Duxford sector as part of No. 12 Group. Led by Squadron Leader P. C. Pinkham, No. 19 Squadron had 15 aircraft on this date, 11 of them serviceable.

RAF Fighter Command

1941–43

With the Battle of Britain won but the Germans still bombing British cities in the 'Blitz', the early weeks of 1941 saw Fighter Command go over to the offensive, trying several types of tactic to pin German air forces in Western Europe and draw them into the air for a battle of attrition.

From the start of 1941, Fighter Command began an attempt to wrest control of the air from the Germans over north-west Europe. By May 1941, the squadrons on the main air bases had been grouped into fighter wings under a single experienced leader.

Several short-penetration fighter tactics were tested in an effort to draw the Luftwaffe into a war of attrition, and to pin German fighter forces in north-west Europe, especially after the June 1941 German invasion of the USSR.

The fighter best suited to the task was the Supermarine Spitfire, which was used in large numbers with medium bomber forces whose task was to attack the targets that the Germans fighters would seek to protect. In general, though, the Germans responded only when the situation favoured them, and the general inaccuracy of the bombing meant that the British attacks were seldom effective.

Thus the results of such operations in 1941 were very mixed. One major problem was the short range of the Spitfire, which had been designed as a limited-endurance interceptor and therefore could not

SELECTED FIGHTER COMMAND HURRICANE SQUADRONS (1941)

Unit	Type	Base
No. 71 Squadron	Hurricane	Church Fenton
No. 91 Squadron	Hurricane	Hawkinge
No. 92 Squadron	Hurricane	Biggin Hill
No. 116 Squadron	Hurricane	Hatfield
No. 121 Squadron	Hurricane	Kirton-in-Lindsey
No. 128 Squadron	Hurricane	Hastings
No. 133 Squadron	Hurricane	Coltishall
No. 136 Squadron	Hurricane	Kirton-in-Lindsey
No. 181 Squadron	Hurricane	Duxford
No. 182 Squadron	Hurricane	Martlesham Heath
No. 239 Squadron	Hurricane	Hatfield
No. 247 Squadron	Hurricane	Roborough
No. 255 Squadron	Hurricane	Kirton-in-Lindsey
No. 306 Squadron	Hurricane	Church Fenton
No. 308 Squadron	Hurricane	Speke
No. 315 Squadron	Hurricane	Acklington
No. 316 Squadron	Hurricane	Pembrey
No. 317 Squadron	Hurricane	Acklington
No. 331 Squadron	Hurricane	Catterick
No. 401 Squadron	Hurricane	Digby
No. 402 Squadron	Hurricane	Digby

Ground attack Hurricane

Soon hard-pressed in the interceptor role, the Hurricane was converted into a ground attack aircraft, armed with Vickers 40mm (1.5in) cannons. The first of this type to enter service flew in the Western Desert in mid-1942.

ORGANIZATION

Fighter Command
January 1941

- No. 9 Group (5 squadrons)
- No. 10 Group (13 squadrons)
- No. 11 Group (26 squadrons)
- No. 12 Group (15 squadrons)
- No. 13 Group (12 squadrons)
- No. 14 Group (5 squadrons)

remain long over France. In these circumstances, the *Luftwaffe* was able to reduce its strength in France to just two *Jadgdeschwadern* of some 180 fighters, although these units were the very capable JG 2 and JG 26. The situation in which British fighters were operating at short range over home territory, with all its recovery and repair facilities, was now entirely reversed in the favour of the Germans. In 1941, therefore, Fighter Command lost some 530 aircraft and the Germans only 236, although British pilots claimed 731. As 1941 ended, the Focke-Wulf Fw 190 appeared as a fighter superior to the Spitfire Mk V, and 1942 and 1943 were therefore harder years in which the offensive was pressed, regardless of losses, to show support for the USSR. Over the UK, meanwhile, improved British night-fighters made the Blitz very costly, and in May it was called off to allow German concentration of effort against the USSR.

Offensive sweeps

1941–43

Fighter Command initially attempted to draw the *Luftwaffe* into the air with massed fighter sweeps, known as 'Rhubarb' operations, but the Germans generally refused to respond. Little more success attended the 'Circus' operations, which were fighter-escorted bomber raids.

THE FIRST TYPE of offensive developed in 1941 by No. 11 Group was the 'Rhubarb', a sweep by the aircraft of one or sometimes two fighter wings. As the sweeps offered little real threat, the Germans largely ignored them. No. 11 Group, therefore, devised the 'Circus', which involved large numbers of fighters escorting a few bombers whose attentions were designed to inflict real damage, and so spur a German reaction. Between January and June 1941, the

ORGANIZATION

Cologne Power Station Attack
12 August 1941

Attack Force (54 Blenheims)	Close Escort Force (one Whirlwind and three Spitfire Squadrons)	Target Support Force (five Spitfire squadrons)	Withdrawal Cover Force (six Spitfire squadrons)

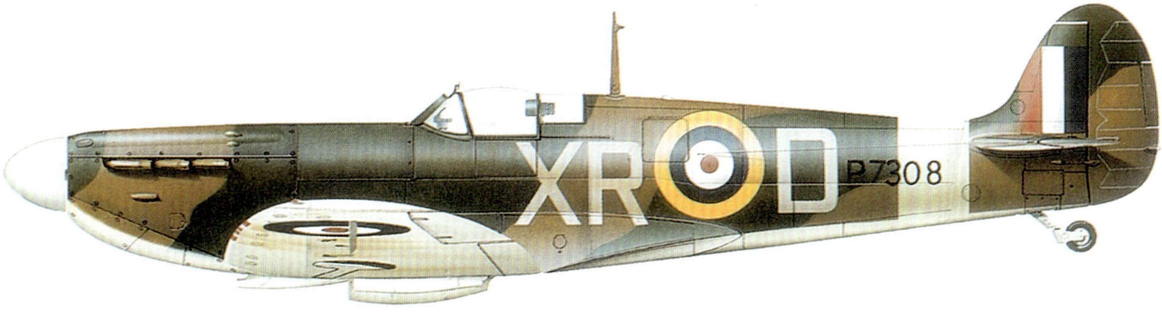

Specifications

Crew: 1
Powerplant: 768kW (1030hp) Rolls-Royce Merlin III liquid-cooled V-12
Maximum speed: 582km/h (362mph)
Range: 636km (395 miles)
Service ceiling: 9725m (31,900ft)
Dimensions: span 11.22m (36ft 10in); length 9.11m (29ft 11in); height 2.69m (8ft 10in)
Weight: 2624kg (5784lb) loaded
Armament: 8 x 7.7mm (0.303in) Browning MGs in wings

Supermarine Spitfire Mk IA

No. 71 'Eagle' Squadron / Fighter Command, 1941

The pilots of this squadron were primarily American volunteers, and this aeroplane was flown by Pilot Officer W. Dunn.

1941 'Circus' sweep

A typical 'Circus' operation of 1941 might involve 120 aircraft in the form of 12 Bristol Blenheim light bombers (one squadron) escorted and covered by 108 Supermarine Spitfire fighters (nine squadrons each comprising four flights of three aircraft). While part of the Spitfire force provided escort for the vulnerable Blenheim bombers, the rest of the force operated semi-independently at higher altitude, ready to pounce on any German fighters that rose to the challenge.

Blenheim squadron

Escort squadron (1 of 9)

Flight 1 **Flight 2** **Flight 3** **Flight 4**

Circuses involved 190 bomber sorties, while the Rhubarbs and Circuses combined involved 2700 fighter sorties.

During these, Fighter Command lost 51 pilots, and claimed the destruction of 44 German aircraft, though the total was in fact somewhat lower. The number of Circuses was increased in the second half of 1941 and into 1942–43. Even though the British had introduced the improved Spitfire Mk V, the Germans had responded initially with the fine Messerschmitt Bf 109F and then in September 1941 with the excellent Focke-Wulf Fw 190A, the latter decidedly better than the Spitfire Mk V. Moreover, the German radar system was now better than the British system had been in 1940, and this made it almost impossible for the Circuses to achieve surprise. Even so, Leigh-Mallory claimed that between 14 June and 3 September 1941, No. 11 Group had destroyed 437 German fighters, with another 182 claimed as probables.

The *Luftwaffe* had no more than about 260 (200 serviceable) single-engined fighters in France and the Low Countries, so the real figures were, in fact, 128 lost and 76 damaged. Fighter Command itself lost 194 pilots during this period. For the period between 14 June and 31 December, Fighter Command claimed 731 'kills' for the loss of 411 aircraft, while the actual German losses were just 154.

Better used in other theatres?

Fighter Command retained 75 day fighter squadrons in the UK in the later part of 1941. The same basic tactical situation prevailed during 1942 and 1943, and is perhaps typified by the amphibious 'reconnaissance in force' to Dieppe on 19 August 1942. The Canadian 2nd Division and supporting elements assaulted under cover of the guns of eight destroyers and Hurricanes firing 20mm (0.79in) cannon, and the operation was a disaster despite an air umbrella of 70 squadrons, including 61 of fighters. Leigh-Mallory claimed 43 German bombers and 49 fighters destroyed, 10 bombers and 29 fighters probably destroyed, and 56 bombers and 84 fighters damaged. In fact, the German losses were 25 bombers and 23 fighters destroyed, and 16 bombers and eight fighters damaged. The RAF lost 106 aircraft, of which 88 were fighters.

▲ **Fighter-bomber**

Loaded with a pair of 114kg (250lb) bombs under its wing, this Hawker Hurricane Mk IIB of No. 402 Squadron, RCAF, prepares for a cross-Channel sortie in 1941.

Specifications

Crew: 1
Powerplant: 876kW (1175hp) Rolls-Royce Merlin XII liquid-cooled V-12
Maximum speed: 570km/h (354mph)
Range: 636km (395 miles)
Service ceiling: 11,457m (37,600ft)
Dimensions: span 11.22m (36ft 10in); length 9.11m (29ft 11in); height 2.69m (8ft 10in)
Weight: 2803kg (6172lb) loaded
Armament: 8 x 7.7mm (0.303in) Browning MGs in wings

▲ **Supermarine Spitfire Mk II**

No. 312 Squadron / RAF Fighter Command, 1941

The aircraft of No. 312 Squadron were flown by Czechoslovakian pilots who had escaped from their homeland at the time of the German takeover in 1939, and this aeroplane was flown by Lieutenant Vybiral. Many of the Czech pilots had gained combat experience with the French Air Force in May and June 1940.

Specifications

Crew: 1
Powerplant: 860kW (1150hp) Allison piston engine
Maximum speed: 580km/h (360mph)
Range: 1100km (650 miles)
Service ceiling: 8800m (29,000ft)
Dimensions: span 11.38m (37ft 4in); length 9.66m (31ft 8in); height 3.76m (12ft 4in)
Weight: 3760kg (8280lb) loaded
Armament: 6 x 12.7mm (0.5in) M2 Browning MGs

▲ Curtiss Tomahawk Mk IIB

No. 414 Squadron / RAF, Croydon, early 1941

In theory, the Tomahawk offered adequate capabilities and was readily available from the USA, but experience soon revealed the type's complete unsuitability for use against a technically sophisticated opponent such as the *Luftwaffe*.

Specifications

Crew: 1
Powerplant: 860kW (1150hp)
Maximum speed: 580km/h (360mph)
Range: 1100km (650 miles)
Service ceiling: 8800m (29,000ft)
Dimensions: span 11.38m (37ft 4in); length 9.66m (31ft 8in); height 3.76m (12ft 4in)
Weight: 3760kg (8280lb) loaded
Armament: 6 x 12.7mm (0.5in) M2 Browning MGs

▲ Curtiss Tomahawk Mk IB

No. 400 Squadron / Royal Canadian Air Force / Odiham, late 1941

After working up in the UK, most Tomahawk squadrons were diverted to secondary theatres, such as those in North Africa, and served in the ground-attack role, a role in which they faced mainly inferior Italian aircraft and AA fire.

Specifications

Crew: 1
Powerplant: 895kW (1200hp) Allison V-1710-85 liquid-cooled V-12
Maximum speed: 605km/h (376mph)
Range: 483km (300 miles)
Service ceiling: 11,665m (38,270ft)
Dimensions: span 10.36m (34ft 0in); length 9.195m (30ft 2in); height 3.785m (12ft 5in)
Weight: 3992kg (8800lb) loaded
Armament: 1 x 37mm (1.46in) M4 cannon, 2 x 12.7mm (0.5in) Browning MGs, 4 x 7.62mm (0.3in) Browning MGs

▲ Bell Airacobra Mk I

No. 601 (County of London) Squadron / Duxford, late 1941–early 1942

The British ordered 675 examples of the Airacobra, but deliveries were halted after the arrival of only 80 aircraft. Some of these went to No. 601 Squadron, which was the only British squadron to be equipped with the type. The Airacobra remained in first-line British service only between October and December 1941.

Fighter-bombers over the Channel

JANUARY–JULY 1942

The British learned much from their first year of sweeps across the English Channel, but also suffered heavy losses. The Germans maintained the technical ascendancy, and the British had to rethink their tactical concepts in the first stages of 1942.

DESPITE EARLIER losses and the need for reinforcements deploying to overseas theatres, the RAF relaunched its offensive operations into northern Europe during March 1942. On 13 March, Circuses were resumed and 'Ramrods' were authorized, the latter focusing on damaging worthwhile land targets, not seeking to draw up and defeat the German fighter arm. Also resumed were Rhubarbs and 'Roadsteads', the last directed at German coastal shipping. At this time, Fighter Command had just over 57 squadrons of Spitfire Mk V fighters (1130 aircraft), in addition to several squadrons equipped with Hurricane Mk IIs, Beaufighters, Whirlwinds and Havocs.

New offensive

The renewed offensive started on 8 March when No. 2 (Bomber) Group's new Boston medium bombers were involved in Circuses against Comines and Abbeville against light opposition. Other Circuses and Ramrods followed, No. 11 Group selecting targets within the combat radius of the Spitfire Mk V. Few were airfields, favoured targets being marshalling yards, chemical plants and small factories. At the same time, No. 10 Group flew the occasional operation to north-western France, while No. 12 Group limited its efforts to Rhubarbs and 'Rodeos' over the southern Netherlands.

During the first month, Fighter Command claimed 53 aircraft downed, 27 probables and 36 damaged, and lost 32 of its own aircraft together with 28 pilots killed or missing. The actual German losses were 12 aircraft. However, April 1942 was a critical month for Fighter Command. In northern France, moves had already been made to strengthen

ORGANIZATION

Circus attack on Hazebrouck
13 March 1942

Bomber Force (11 Douglas Bostons from (Nos 88 & 226 Squadrons)	Close Escort Force (4 x Spitfire squadrons from Northolt and Hornchurch Wings)	1st Fighter Echelon (6 x Spitfire squadrons from Kenley and Biggin Hill Wing)	2nd Fighter Echelon (4 x Spitfire squadrons from Tangmere Wing)

Specifications

Crew: 1
Powerplant: 860kW (1150hp)
Maximum speed: 580km/h (360mph)
Range: 1100km (650 miles)
Service ceiling: 8800m (29,000ft)
Dimensions: span 11.38m (37ft 4in); length 9.66m (31ft 8in); height 3.76m (12ft 4in)
Weight: 3760kg (8280lb) loaded
Armament: 6 x 12.7mm (0.5in) M2 Browning MGs

▲ Curtiss Tomahawk Mk I

No. 400 Squadron / RAF, late 1941

Another US warplane ordered in large numbers, the Tomahawk proved inadequate at all but low levels for lack of adequate supercharging, and was allocated to 16 home-based tactical reconnaissance squadrons, seeing only limited use before being replaced by North American Mustangs.

the defence and, on a limited basis, go over to the offensive with specially created *Jagdbomberstaffeln* (fighter-bomber squadrons) supplementing some initial units that had already undertaken successful tip-and-run attacks on British south coast towns, convoys and radar stations. During the 'Circus' operations of April 1942, claims of 40 destroyed, 32 probable and 63 damaged were made for the loss of 81 aircraft, but the losses of *Jagdgeschwader* 2 (JG 2) and JG 26 were in fact only 21 fighters.

July 1942 saw the introduction of the so-called 'Mass Rhubarb', a tactic in which 120 or more Spitfires roamed northern France at tree-top height to attack targets of opportunity. The German anti-aircraft guns exacted a heavy toll. In the six months to the end of July 1942 on the Channel Front, Fighter Command lost 335 fighters, whereas the Germans lost only 84 fighters.

FIGHTER COMMAND WHIRLWIND SQUADRONS, 1941		
Unit	Type	Base
No. 137 Squadron	Whirlwind	Charmy Down
No. 263 Squadron	Whirlwind	Charmy Down

▲ Hawker Hurricane Mk IIC

No. 1 Squadron / RAF Fighter Command, 1942

By early in 1942, the Hurricane no longer possessed the performance to operate in the pure fighter role, but the Hurricane Mk IIC fighter-bomber remained in limited service from British bases. It was used in larger numbers and with greater success in the North Africa and Burma theatres.

Specifications

Crew: 1
Powerplant: 954kW (1280hp) Rolls-Royce Merlin XX liquid-cooled V-12
Maximum speed: 529km/h (329mph)
Range: 1480km (920 miles)
Service ceiling: 10,850m (35,600ft)
Dimensions: span 12.19m (40ft 0in); length 9.81m (32ft 2.25in); height 3.98m (13ft 1in)
Weight: 3629kg (8044lb) loaded
Armament: 4 x 20mm (0.79in) Hispano cannon and up to 544kg (1000lb) of bombs or rocket projectiles

Specifications

Crew: 1
Powerplant: 1685kW (2260hp)
Maximum speed: 650km/h (405mph)
Range: 980km (610 miles)
Service ceiling: 10,400m (34,000ft)
Dimensions: span 12.67m (41ft 7in); length 9.73m (31ft 11in); height 4.66m (15ft 4in)
Weight: 5170kg (11,400lb) loaded
Armament: 4 x 20mm (0.8in) Hispano-Suiza HS.404 cannons, 2 x 454 kg (1000lb) bombs

▲ Hawker Typhoon Mk IB

No. 3 Squadron / RAF Fighter Command, May 1943

Conceived as an interceptor and introduced in September 1941, the Typhoon failed in its intended role, but possessed excellent speed at low altitude. But the type was important in defeating the German 'tip-and-run' fighter-bomber raids on the south coast of the UK, and then became a superb ground-attack fighter.

Specifications

Crew: 1
Powerplant: 858kW (1150hp) Allison V-1710-39 V12 engine
Maximum speed: 690km/h (430mph)
Range: 1215km (755 miles)
Service ceiling: 12,649m (41,500ft)
Dimensions: span 11.27m (37ft); length 9.84m (32ft 4in); height 4.15m (13ft 8in)
Weight: 4173kg (9200lb) loaded
Armament: 4 x 12.7mm (0.5in) and 4 x 7.62mm (0.3in) Browning MGs

▲ **North American Mustang Mk I**

No. 26 Squadron / RAF Army Cooperation Command, Gatwick, mid–late 1942

Powered by an Allison V-1710 V-2 engine of the type also used in the Tomahawk, the Mustang Mk I also possessed inferior performance at altitude, and after entering service in April 1942 was used in the low-altitude reconnaissance fighter role.

Coastal defence

1941–43

While the bulk of the RAF's fighter strength was devoted to offensive operations over north-west Europe, the Germans still possessed the capability to make individually small but cumulatively significant attacks on British ports and coastal convoys. Fighter Command had to respond.

THROUGHOUT WORLD WAR II, the UK remained reliant on her ports. At the ports arrived convoys delivering vital food, raw supplies, men and weapons from her empire, the United States and other countries; and from them men, equipment and other exports went off to different theatres and, ultimately, northwest France for the Allied invasion of Normandy in June 1944. Thus the major and even minor ports around the country's long coastline were inevitable objectives for German attack, and both oceanic and coastal convoys also proved to be very important targets for the *Luftwaffe* and *Kreigsmarine*.

German raids

Much of the work against ports and coastal convoys was undertaken by submarines and coastal craft with torpedoes and mines, while submarines and long-

COASTAL COMMAND BEAUFIGHTER SQUADRONS (1941–45)

Unit	Date
No. 143 Squadron	June 1941 – October 1944
No. 144 Squadron	May 1943 – May 1945
No. 235 Squadron	December 1941 – June 1944
No. 236 Squadron	October 1941 – May 1945
No. 248 Squadron	July 1941 – January 1944
No. 252 Squadron	December 1940 – December 1946
No. 254 Squadron	June 1942 – October 1946
No. 404 Squadron (RCAF)	September 1942 – April 1945
No. 406 Squadron (RCAF)	June 1941 – August 1944
No. 455 Squadron (RAAF)	December 1943 – May 1945
No. 498 Squadron (RNZAF)	November 1943 – August 1945

ORGANIZATION

RAF Coastal Command
1941

- Fighter escort and patrol squadrons
- Reconnaissance squadrons
- Air/sea rescue squadrons
- Flying boat squadrons

range aircraft targeted oceanic convoys with torpedoes and bombs. Yet there was also a part to be played by the *Luftwaffe*'s fighter-bombers and coastal aircraft, the former delivering pinpoint attacks that could secure limited useful physical results at times, but also served on a larger scale to tie down British forces and keep the British defences off balance. German coastal aircraft served by laying mines in the estuaries of British rivers and other natural chokepoints off British ports.

The German air activity was normally undertaken at low level to ensure that the fighter-bombers and minelayers suffered the minimum possibility of radar-directed interception. This tactic in turn required the British to operate a system of standing

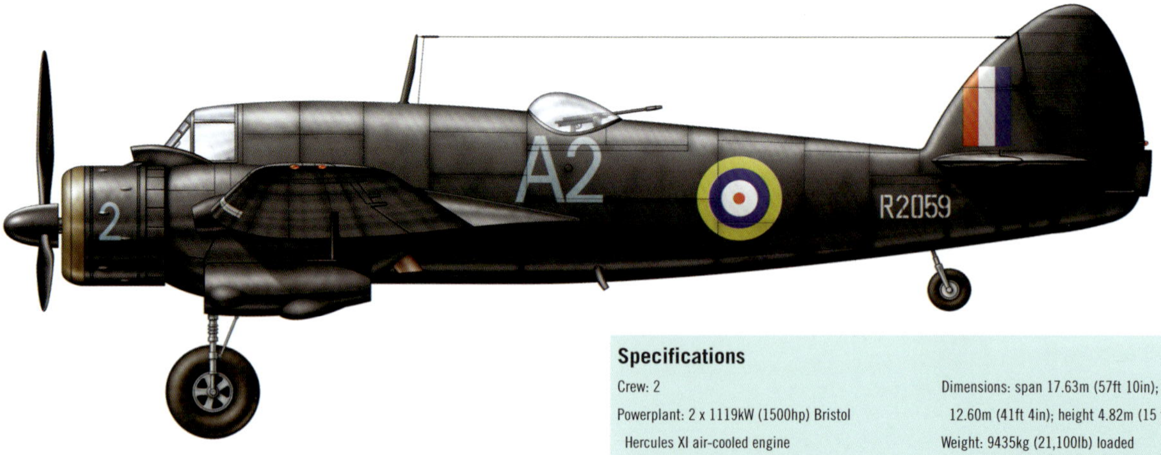

▲ **Bristol Beaufighter Mk IF**

RAF Fighter Command, UK, 1941

Carrying antennae on its nose and the leading edges of its outer wing panels, the Beaufighter Mk IF was the world's first effective radar-equipped night-fighter.

Specifications

Crew: 2
Powerplant: 2 x 1119kW (1500hp) Bristol Hercules XI air-cooled engine
Maximum speed: 492km/h (306mph)
Range: 2414km (1500 miles)
Service ceiling: 8810m (28,900ft)
Dimensions: span 17.63m (57ft 10in); length 12.60m (41ft 4in); height 4.82m (15 ft 10in)
Weight: 9435kg (21,100lb) loaded
Armament: 4 x 20mm (0.79in) cannon and 6 x 7.7mm (0.303in) MGs in the wings, and 1 x 7.7mm (0.303in) Vicker rearward-firing MG in dorsal turret

▲ **Boulton Paul Defiant Mk II**

No. 151 Squadron / RAF Fighter Command, 1941

Though never a truly effective fighter, the Defiant was better as a night-fighter than as a conventional day-fighter.

Specifications

Crew: 2
Powerplant: 954kW (1280hp) Rolls-Royce Merlin XX liquid-cooled V-12
Maximum speed: 504km/h (313mph)
Range: 748km (465 miles)
Service ceiling: 9250m (30,350ft)
Dimensions: span 11.99m (39ft 4in); length 10.77m (35ft 4in); height 3.71m (12ft 2in)
Weight: 3821kg (8424lb) loaded
Armament: 4 x 7.7mm (0.303in) Browning MGs in the power-operated dorsal turret

patrols from the many air bases, large and small, located round the coast of the UK. With the aid of such radar direction as was possible, and visual sightings by shore-based observers and the personnel of coastal shipping, at least a modicum of a defence could be made, and sufficient losses inflicted on the Germans to deter their efforts.

Along the south coast of England, 'tip-and-run' raids made by high-performance German fighter aircraft, such as the manouevrable Focke-Wulf Fw 190A and fast-turning Messerschmitt Bf 109F, could only be countered by British fighters of equal performance, typically the latest variants of the redoubtable Supermarine Spitfire and then the Hawker Typhoon.

The Typhoon was known for its excellent turn of low-altitude speed and the devastating armament of four 20mm (0.79in) cannon, which could wreck almost any aircraft with only a half-second burst. These aircraft had effectively halted the main

Specifications

Crew: 1
Powerplant: 954kW (1280hp) Rolls-Royce Merlin XX liquid-cooled V-12
Maximum speed: 529km/h (329mph)
Range: 1480km (920 miles)
Service ceiling: 10,850m (35,600ft)
Dimensions: span 12.19m (40ft 0in); length 9.81m (32ft 2.25in); height 3.98m (13ft 1in)
Weight: about 3649kg (8044lb) loaded
Armament: 12 x 7.7mm (0.303in) Browning MGs

Hawker Hurricane Mk IIA

No. 253 Squadron / RAF Fighter Command, 1941

The all-black finish, alleviated only by national and unit markings, was designed to make the Hurricane 'invisible' in the night sky.

Specifications

Crew: 1
Powerplant: 2 x 659kW (885hp) Peregrine engines
Maximum speed: 580km/h (360mph)
Range: 1300km (808 miles)
Service ceiling: 9240m (30,315ft)
Dimensions: span 13.72m (45ft); length 9.83m (32ft 3in); height 3.53m (11ft 7in)
Weight: 4697kg (10,356lb) loaded
Armament: 4 x Hispano 20mm (0.8in) cannon in nose (60 rounds per gun, 240 rounds total)

Westland Whirlwind

No. 263 Squadron / RAF Fighter Command, west of England, 1942

No. 263 Squadron was one of only two squadrons equipped with the Whirlwind twin-engined long-range fighter and fighter-bomber. With its good range and considerable firepower, the Whirlwind was well suited to the coastal role, in defence as well as offence.

German effort by the end of 1942. Along the rest of the UK's coast, where the threat was posed by lower-performance aircraft such as the Heinkel He 115 twin-engined floatplane and Messerschmitt Bf 110 twin-engined multi-role warplane, older aircraft were used since these were adequate to the task and were no longer needed for cutting-edge operations.

So there was a steady flow of fighter patrols, supported by other aircraft, over the coasts and coastal waters of the UK during the mid years of the war. These patrols served a useful primary purpose, and also provided a facility for the locating and marking of men from downed aircraft or sunken ships, allowing air or sea rescue craft to be called in and directed to the right spot.

The patrols also provided useful flying and navigation training for inexperienced pilots and those recovering from wounds or being rested for tiredness.

Fleet Air Arm

1939–43

Returning to the full control of the Royal Navy after years in the technical doldrums under the control of the RAF, the Fleet Air Arm was poorly equipped at the technical level for the type of warfare it would have to wage between 1939 and 1945, but was full of ideas and spirit.

THE ROYAL NAVAL AIR SERVICE was merged with the Royal Flying Corps (RFC) on 1 April 1918 to create the Royal Air Force. The RAF kept control of the Naval Air Branch until 1937, when it was returned to Admiralty control.

At the start of World War II, it had become the Fleet Air Arm (FAA) and comprised 20 squadrons with only 232 aircraft, most of them obsolescent and none up to the standards of modern land-based warplanes.

ORGANIZATION

Fleet Air Arm
1940

Carrierborne fighter squadrons	Home-based fighter squadrons	Overseas fighter squadrons	Fighter training establishments

Naval fighters

The naval fighters of this time were the Blackburn Roc and then the Fairey Fulmar, and the FAA then progressed through the Sea Hurricane and Supermarine Seafire to US types such as the Grumman Hellcat and Vought Corsair. By the end of World War II, when the FAA had been deployed all over the world, the FAA was operating from 59 aircraft carriers (fleet, light and escort) and 56 air stations, and had some 3700 aircraft and 72,000 men.

FLEET AIR ARM ROC SQUADRONS

Unit	Type	Base
No. 801 Squadron	Roc	Hatston
No. 803 Squadron	Roc	n/a
No. 806 Squadron	Roc	Eastleigh
No. 759 Squadron	Roc	n/a
No. 760 Squadron	Roc	n/a
No. 769 Squadron	Roc	n/a
No. 772 Squadron	Roc	n/a
No. 773 Squadron	Roc	n/a

SELECTED FLEET AIR ARM FULMAR SQUADRONS

Unit	Type	Base
No. 800X Squadron	Fulmar	HMS *Furious*
No. 800Y Squadron	Fulmar	HMS *Argus*
No. 800Z Squadron	Fulmar	HMS *Victorious*
No. 803 Squadron	Fulmar	India & Ceylon
No. 804 Squadron	Fulmar	HMS *Eagle*
No. 805 Squadron	Fulmar	Egypt & Crete
No. 806 Squadron	Fulmar	HMS *Illustrious*

▲ Blackburn Skua Mk II

No. 5 Maintenance Unit, Kemble, 1939

The Skua Mk II was the production variant of the Skua Mk I prototype. Through designed as a carrierborne dive-bomber, the Skua doubled as a naval fighter, but was ineffective in each role and withdrawn from first-line service in 1941.

Specifications

Crew: 2
Powerplant: 664kW (890hp) Bristol Perseus XII radial engine
Maximum speed: 362km/h (225mph)
Range: 1223km (760 miles)
Service ceiling: 6160m (20,200ft)
Dimensions: span 14.07m (46ft 2in); length 10.85m (35ft 7in); height 3.81m (12ft 6in)
Weight: 3732kg (8228lb) loaded
Armament: 4 x 7.7mm (0.303in) MGs in wings; Lewis rear gun; 1 x 227kg (500lb) bomb beneath fusilage

▲ Blackburn Roc

No. 2 AACU / Fleet Air Arm

Essentially the naval counterpart of the land-based Boulton Paul Defiant, the Roc was based on the turret fighter concept and suffered the same basic failings as the Defiant. These were the weight of the power-operated turret, a poor power/weight ratio, and the lack of agility and fixed forward-firing armament.

Specifications

Crew: 2
Powerplant: 675kW (905hp) Bristol Perseus XII radial engine
Maximum speed: 359km/h (223mph)
Range: 1304km (810 miles)
Service ceiling: 5485m (18,000ft)
Dimensions: span 14.02m (46ft); length 10.85m (35ft 7in); height 3.68m (12ft)
Weight: 3606kg (7950lb) loaded
Armament: 4 x 7.7mm (0.303in) MGs in dorsal turret

▲ **Gloster Sea Gladiator Mk I**

RAF service, June 1940

The Sea Gladiator Mk I was the full-standard carrierborne fighter derived closely from the Gladiator, which was the UK's last land-based biplane fighter.

Specifications

Crew: 1
Powerplant: 619kW (830hp) Bristol Mercury VIIIAS air-cooled 9-cylinder radial
Maximum speed: 407km/h (253mph)
Range: 684km (425 miles)
Service ceiling: 9845m (32,300ft)
Dimensions: span 9.83m (32ft 3in); length 8.36m (27ft 5in); height 3.52m (11ft 7in)
Weight: 2272kg (5020lb) loaded
Armament: 4 x 7.7mm (0.303in) Browning MGs

Early war

1939–40

On the outbreak of World War II, the Fleet Air Arm was small and possessed only indifferent aircraft. Even so, imbued with the offensive spirit of the Royal Navy, the FAA's squadrons were soon involved in active air operations in many parts of the world.

DESPITE THEIR SMALL numbers, the squadrons of the FAA were able to fulfil an important role early in World War II. On 1 September 1939, the FAA had 232 aircraft, and of these most were inferior to their RAF counterparts. The most modern type in the fighter, dive-bomber and fighter-reconnaissance role was the Blackburn Skua, operational with Nos 800, 801, 803 and 806 Squadrons, which had just 36 aircraft. The primary attacker was the Fairey Swordfish Mk I torpedo and reconnaissance biplane, of which 140 were in service, with the 457mm (18in) torpedo as their primary weapon. The rest of the FAA comprised either shipborne or shore-based spotter and reconnaissance types such as the Supermarine Walrus and Fairey Seafox.

From the beginning of hostilities, the FAA undertook anti-submarine patrols with HMS *Ark Royal*, *Hermes* and *Courageous* in the North-West and

ORGANIZATION

FAA, Royal Navy
1 September 1939

Ark Royal & *Furious* (Home Fleet)	*Courageous* & *Hermes* (Channel Force)	*Glorious* (Mediterranean Fleet)	*Eagle* (China Station)

FAA CARRIERBORNE AIRCRAFT STRENGTHS (1939)

Ship	Aircraft type	Strength
Eagle	Swordfish	18
Hermes	Swordfish	9
Furious	Swordfish, Skua & Roc	18, 8 & 4
Courageous	Swordfish	24
Glorious	Swordfish & Sea Gladiator	36 &12
Ark Royal	Swordfish & Skua	42 & 18

▲ **Early Martlet**

This Martlet Mk I was one of the French Cyclone-powered Martlets delivered in September 1940.

South-West Approaches under the Home Fleet, and with Nos 800, 803, 810, 820, 821, 822 and 814 Squadrons embarked. HMS *Furious*, *Glorious* and *Eagle* were in the Atlantic, off Aden and in the East Indies respectively, and early operations included patrols off the Cape Verde Islands and Brazil, and unsuccessful searches for the *Graf Spee* in the South Atlantic.

The shore-based Nos 800 and 804 Squadrons, based at Hatston-Kirkwall, operated against the German Navy off Norway in April 1940. The *Furious* was involved in attacks in Narvikfjord on 10 April, and again towards the end of the month, until relieved by the *Glorious* and *Ark Royal* with Skua, Roc and Sea Gladiator Mk I warplanes, which saw action off Namsos, Bodo and Narvik.

During September and October 1940, the *Furious* launched two strikes on Tromso and Trondheim. With the loss of the *Glorious* in June 1940, the *Ark Royal* was the only carrier left to Home Fleet, with the others either ferrying or in stations in the Mediterranean and Indian Ocean.

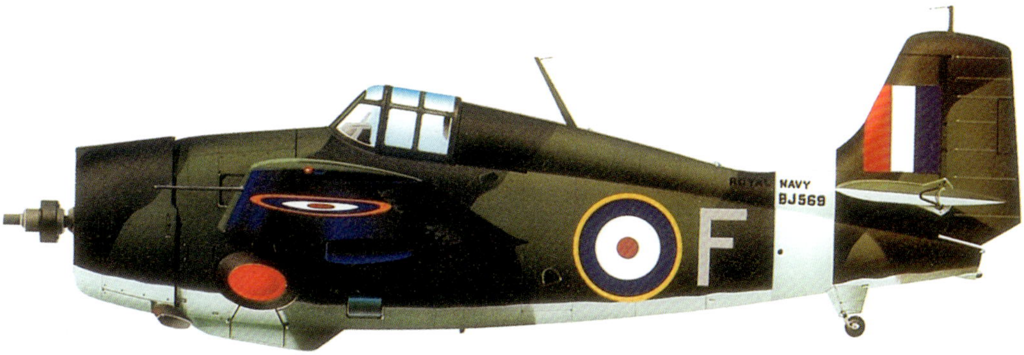

Specifications

Crew: 1
Powerplant: 880kW (1180hp) Pratt & Whitney R-1830-86 engine
Maximum speed: 512km/h (318mph)
Range: 2173km (1350 miles)
Service ceiling: 10,900m (35,700ft)
Dimensions: span 11.58m 38ft); length 8.76m (28ft 8in); height 2.81m (9ft 3in)
Weight: 3607kg (7952lb) loaded
Armament: 4 x 12.7mm (0.5in) MGs in wings; 2 x 113kg (249lb) bombs

▲ **Grumman Martlet Mk I**

No. 804 Squadron / Fleet Air Arm, Hatston, 1940

The Martlet Mk I was the British version of the Grumman F4F-3, and entered service with the FAA in September 1940, initially with No. 804 Squadron based in the Orkney Islands for the defence of the Home Fleet's great anchorage and base in Scapa Flow.

Convoy protection

JANUARY 1941 – MAY 1943

One of the most important tasks undertaken by the Fleet Air Arm was the protection of convoys crossing the Atlantic. Here the FAA's fighters were needed to drive off or shoot down German long-range reconnaissance and bombing aircraft, and keep a constant watch for U-boats.

BY THE END OF 1940, the Allies and many neutral states trading across the Atlantic had lost 1281 ships, some 585 of them to U-boat attack and the others to attacks by aircraft, surface ships both large and small, and mines.

Over the same 16-month period, the Germans had lost only 32 U-boats, none of them to air attack alone. Yet the battle was now beginning to sway in the opposite direction as the British began to field longer-range maritime patrol aircraft, about one-sixth of them fitted with air-to-surface search radar. Larger numbers of better escort vessels were becoming available, and better coordination of air and naval assets was arriving after the creation of the RAF's No. 19 Group.

ORGANIZATION

Joint Royal Navy and Royal Air Force capability
mid-1942

RN escort carriers	RN escort groups	FAA shore-based squadrons	RAF Coastal Command

Convoy attacks

Yet the scale of the task was reflected in a pair of events in February 1941. On 8–12 February, a U-boat vectored Focke Fw 200 long-range aircraft and the cruiser *Admiral Hipper* onto two convoys, which lost 16 out of 26 ships, and then, on 26 February, Fw 200 aircraft were called in by U-boat to sink seven ships of a single convoy and damage another four. In May, Swordfish attack aircraft of the *Ark Royal* were instrumental in damaging and slowing the battleship *Bismarck*, which was then caught and sunk by British surface forces.

So far as convoys were concerned, though, there remained a 'black hole' in the centre of the Atlantic,

Specifications

Crew: 1
Powerplant: 880kW (1180hp) Pratt & Whitney R-1830-86 engine
Maximum speed: 512km/h (318mph)
Range: 2173km (1350 miles)
Service ceiling: 10,900m (35,700ft)
Dimensions: span 11.58m 38ft); length 8.76m (28ft 8in); height 2.81m (9ft 3in)
Weight: 3607kg (7952lb) loaded
Armament: 4 x 12.7mm (0.5in) MGs in wings; 2 x 113kg (249lb) bombs

▲ Grumman Martlet Mk I

No. 804 Squadron / Fleet Air Arm, Hatston, Orkney Islands, early 1941

Used for the air defence of Scapa Flow, this aeroplane was based on the US Navy's F4F-3, and was one of 91 aircraft originally built to a French order. The type had a fixed wing, Wright Cyclone single-row radial engine, and a fixed forward-firing armament of four 12.7mm (0.5in) machine guns.

Specifications

Crew: 2
Powerplant: 805kW (1080hp) Rolls-Royce Merlin VIII V-12 engine
Maximum speed: 398km/h (247mph)
Range: 1336km (830 miles)
Service ceiling: 21,500m (6555ft)
Dimensions: span 14.14m (46ft 4in); length 12.24m (40ft 2in); height 4.27m (14ft)
Weight: 4387kg (9672lb) loaded
Armament: 8 x 7.7mm (0.303in) MGs in wings

▲ **Fairey Fulmar Mk I**

No. 806 Squadron / Fleet Air Arm, 1942

While its layout, powerplant and fixed forward-firing armament were akin to those of contemporary land-based fighters, the Fulmar was rendered a wholly indifferent fighter by the Royal Navy's demand that it have two- rather than one-seat accommodation in a longer and weightier fuselage. The second occupant wielded no armament, so undertook just the observer, navigator and radio operator roles.

too far from the UK and North America to be reached even by long-range aircraft, and here the U-boat was almost impossible to defeat.

Fleet aircraft carriers were too few in number and too valuable to be risked in these circumstances, but an initial answer came with the CAM-ship. This was a merchant vessel fitted with a catapult to launch a Hawker Hurricane fighter, which had to ditch in the sea close to a ship at the end of its single mission, the pilot being recovered. These Hurricanes made life very difficult for the Fw 200 long-range reconnaissance aircraft that called in the U-boat packs, and also spotted for and attacked the U-boats.

The CAM-ship was recognized from the start as only a palliative, and the true solution was found in the escort carrier, which was a merchant vessel adapted for the carriage, launch and recovery of small numbers of fighters and anti-submarine aircraft, increasingly the Grumman Wildcat and Grumman Avenger respectively.

The first such vessel was HMS *Audacity*. The ship survived only to December 1941 before being lost to a torpedo, but the concept had been proved and the subsequent construction of escort carriers in US and British yards allowed the cheap yet effective little aircraft carriers to provide just the type of air escort the Atlantic convoys needed. The Germans responded with measures that included greater use of night cover, heavier anti-aircraft armament and radar warning receivers, to which the Allies in turn responded with better radio direction-finders and improved anti-submarine weapons for ships. For aircraft came heavy fixed forward-firing armament, radar that the Germans found difficult to detect and, for a night-attack capability, the Leigh Light – an underwing searchlight turned on only in the final stages of a radar-guided approach, illuminating the target U-boat for accurate visual attack.

The turning point in the Battle of the Atlantic was May 1943. In this month, the Germans lost 41 U-boats, 38 of them in the Atlantic or in the Bay of Biscay, the latter as U-boats attempted to pass between French ports and their operating areas deep in the Atlantic. There were still successes and reverses for each side, but Germany could no longer sustain U-boat losses at the rate of May 1943.

ROYAL NAVY ESCORT CARRIERS (1941–43)

Class	Displacement	Speed	Air strength
'Audacity' (1)	11,000 tons	15kt	6
'Activity' class (1)	14,250 tons	18kt	11
'Vindex' (2)	16,830 tons	18kt	18
'Campania' (1)	15,970 tons	16kt	18
'Archer' (1)	12,860 tons	16.5kt	16
'Avenger' (3)	15,300 tons	16.5kt	15
'Attacker' (11)	14,170 tons	18.5kt	18–24

RAF Desert Air Force
1940–42

One of the lessons of the unsuccessful British *Battleaxe* offensive of June 1941 was the need of the Allied ground forces for air support that was both better organized and larger in scope and capability. This led to the creation of the Desert Air Force to provide tactical air support.

The Desert Air Force (DAF; later 1st Tactical Air Force) was created in North Africa to provide close air support to the British Eighth Army, and comprised British, Australian, South African and finally US squadrons. Before the creation of the Desert Air Force, several RAF formations operated in North Africa.

By the time of Italy's declaration of war in June 1940, the British air commander in the Middle East, Air Vice Marshal Sir Arthur Longmore, had only 29 squadrons with fewer than 300 aircraft. In 1941,

ORGANIZATION

Air Headquarters Western Desert (Air Vice Marshal A. Coningham)
23 October 1941

- No. 201 Group (Air Commodore L. H. Slatter)
- No. 202 Group (Air Commodore T. W. Elmhirst)
- No. 205 Group (Air Commodore L. L. MacLean)
- No. 206 Group (Air Commodore C. B. Cooke)

WDAF FIGHTER & TACTICAL RECCE SQUADRONS (26 MAY 1942)

Unit	Type	Base
No. 2 Sqn, SAAF	Kittyhawk	Gambut
No. 3 Sqn, RAAF	Kittyhawk	Gambut
No. 4 Sqn, SAAF	Tomahawk	Gambut
No. 5 Sqn, SAAF	Tomahawk	Gambut
No. 15 Sqn, SAAF	Blenheim IVF	Amiriya, Kufra
No. 33 Sqn, RAF	Hurricane	Gambut
No. 40 Sqn, SAAF	Hurricane, Tomahawk	El Adem
No. 73 Sqn, RAF	Hurricane	El Adem
No. 80 Sqn, RAF	Hurricane	Gambut
No. 112 Sqn, RAF	Kittyhawk	Gambut
No. 145 Sqn, RAF	Hurricane	Gambut
No. 208 Sqn, RAF	Hurricane, Tomahawk	El Adem
No. 250 Sqn, RAF	Kittyhawk	Gambut
No. 260 Sqn, RAF	Kittyhawk	Gambut
No. 274 Sqn, RAF	Hurricane	Gambut
No. 450 Sqn, RAAF	Kittyhawk	Gambut

▲ **Curtiss Tomahawk Mk IIB**

No. 112 Squadron / Western Desert Air Force, Sidi Haneich, October 1941

Tomahawk Mk IIB was the designation that the Royal Air Force applied to the British-ordered version of the Hawk 81A-2 ordered by France but delivered to the UK after the fall of France for service with the designation Tomahawk Mk IIA.

Specifications

Crew: 1
Powerplant: 860kW (1150hp)
Maximum speed: 580km/h (360mph)
Range: 1100km (650 miles)
Service ceiling: 8800m (29,000ft)
Dimensions: span 11.38m (37ft 4in); length 9.66m (31ft 8in); height 3.76m (12ft 4in)
Weight: 3760kg (8280lb) loaded
Armament: 6 x 12.7mm (0.5in) M2 Browning MGs in wings

command passed to Air Marshal Arthur Tedder, who reorganized his strength into wings, of which the first was No. 253 Wing for close support. The first command-level formation, formed on 21 October 1941, was the Air Headquarters Western Desert (AHWD), created by revision of No. 204 Group.

Specifications	
Crew: 1	Dimensions: span 9.14m (30ft); length 8.15m (26ft 9in); height 3.10m (10ft 2in)
Powerplant: 477kW (640hp) Rolls-Royce Kestrel VI engine	Weight: 1637kg (3609lb) loaded
Maximum speed: 359km/h (223mph)	Armament: 2 x 7.7mm (.303in) Vickers forward firing MGs
Range: 435km (270 miles)	
Service ceiling: 8990m (29,500ft)	

▲ Hawker Fury Mk II

No. 43 Squadron / South African Air Force

The South African Air Force bought seven of these aircraft in 1936, and at the start of World War II received 24 ex-RAF aircraft. A number of Furys were used against Italian forces in East Africa in 1941.

North Africa

1942–43

After a promising tactical start blunted by the German air units' technical edge, the Desert Air Force matured steadily into an exceptional tactical air arm whose later aircraft and fully fledged tactics set the pattern for the tactical air support in the mainland European campaigns.

THE AHWD INITIALLY OPERATED in three wings as Nos 258 and 269 Wings over the front and No. 262 Wing over the Nile delta. On 20 January 1942, the AHWD became Air Headquarters Libya, but on 3 February reverted to AHWD.

On 31 January 1943 command was assumed by Air Vice Marshal H. Broadhurst, and on 10 July 1943 the force was renamed as the DAF, which was subordinated to the North-West African (later Mediterranean and 1st) Tactical Air Force.

Given Fighter Command's priority, the North African force was generally equipped with obsolescent aircraft types such as the Gloster Gladiator, but nonetheless performed well against the Italian air force. As the direct threat to the UK declined, North Africa started to receive more modern aircraft, including the Hawker Hurricane, and the US-built Curtiss Tomahawk/Kittyhawk

ORGANIZATION

North-West African Tactical Air Force (Air Marshal Sir Arthur Coningham)
mid-April 1943

- North African Tactical Bomber Force (Air Vice Marshal Laurence Sinclair)
- Desert Air Force (Air Vice Marshal H. Broadhurst)

FIGHTER WINGS OF NO. 211 GROUP, DAF (MID-APRIL 1943)		
Unit & Squadrons	**Type**	**Strength**
No. 7 Wing, SAAF Nos 2, 4, 5	Kittyhawk	63
US 57th Fighter Group 64th, 65th, 66th & 314th	Kittyhawk	80
US 79th Fighter Group 85th, 86th, 87th & 316th	Kittyhawk	80
No. 239 Wing, RAF Nos 3 & 450 RAAF, & Nos 112, 250, 260 & 450 RAF	Kittyhawk	105
No. 244 Wing, RAF No. 1 SAAF, No. 417 RCAF, & Nos 6, 92, 145 & 601 RAF	Hurricane & Spitfire	126
No. 285 Wing, RAF No. 40 SAAF, & No. 73 RAF	Hurricane & Spitfire	39

series, which was not suitable for operations in northern Europe.

Numbers advantage

The DAF always outnumbered its Axis opponents, and devoted its primary effort to tactical support and long-range interdiction. These pitted the DAF's warplanes against the technically superior Messerschmitt Bf 109F fighters of the *Luftwaffe*'s *Jagdgeschwader* 27, whose pilots in general had the advantage of altitude and surprise over the comparatively slow and low-altitude DAF fighters, which suffered heavy losses.

During 1942, the DAF revised its tactics and received more modern aircraft, the latter including the Supermarine Spitfire from August 1942 for use in the air superiority role, and this helped to tun the tide in the DAF's favour. The DAF adapted the German concept of fighter-bombers controlled by forward air controllers using radio and attached to frontline ground units. The DAF improved the concept by introducing 'cab ranks' of fighter-bombers waiting to be directed at a specific tactical targets.

In this way, the DAF provided vital and decisive air support to the Eighth Army until the end of the war, fighting through Egypt, Libya, Tunisia, Sicily and mainland Italy. The tactics that had proved successful in the latter part of the North African campaign were also adopted with even greater success during the invasion of Europe in 1944.

Commonweath flyers

The major Commonwealth contributor to the DAF was South Africa, which provided more than 12 squadrons. North Africa was the South Africans' primary theatre of service, as their government had decided that no South African formations should fight outside Africa. The Australian element included No. 3 Squadron, which reached North Africa late in

▲ **Curtiss Kittyhawk Mk I**

No. 112 Squadron / Western Desert Air Force, 1942

Like the closely related Tomahawk, the Kittyhawk was a member of the tactical fighter family known to the US Army Air Forces as the P-40. The type operated almost exclusively in the low-level fighter-bomber role with weapons such as a 227kg (500lb) bomb under the fuselage.

Specifications

Crew: 1

Powerplant: 895kW (1200hp) Allison piston engine

Maximum speed: 563km/h (350mph)

Range: 1738km (1080 miles)

Service ceiling: 9450m (31,000ft)

Dimensions: span 11.36m (37ft 4in); length 10.16m (33ft 4in); height 3.76m (12ft 4in)

Weight: 3511kg (7740lb) loaded

Armament: 4 x 12.7mm (0.5in) Browning MGs in wings

1940 and served with the DAF until the closing stages of the war in Europe. Many foreign personnel also flew in RAF squadrons, some of the most numerous being Polish pilots and other aircrew.

The US Army Air Forces (USAAF) provided more strength: the 57th and 79th Fighter Groups, which flew the P-40, and one medium bomber group, which flew the North American B-25, served with the DAF from mid-1942 until they were absorbed into the new US Ninth Air Force.

The Western Desert Air Force had about 1000 warplanes by late 1941. By the time of the second battle of El Alamein late in 1942, the DAF had more than 1500, more than double the number of Axis aircraft.

BRITISH SQUADRONS IN THE MEDITERRANEAN THEATRE

Date	Fighter	Light bomber
September 1939	6	8
July 1940	8	9
December 1941	29	10
December 1942	47	17
December 1943	49	13
September 1944	41	7
March 1945	33	6

Specifications

Crew: 1

Powerplant: 988kW (1325hp) Allison V-1710-87 liquid-cooled piston V12 engine

Maximum speed: 590km/h (365mph)

Range: 885km (550 miles)

Service ceiling: 7650m (25,100ft)

Dimensions: span 11.28m (37ft 1in); length 9.83m (32ft 3in); height 3.71m (12ft 2in)

Weight: 4535kg (10,000lb) loaded

Armament: 6 x 12.7mm (0.50in) M2 Browning MGs in wings

▲ **North American A-36A**

No. 1437 Strategic Reconnaissance Flight

Fast at low level and possessing good range, the A-36 was well suited to the reconnaissance role. This aircraft was used in Tunisia early in 1943.

Specifications

Crew: 1

Powerplant: 860kW (1150hp)

Maximum speed: 580km/h (360mph)

Range: 1100km (650 miles)

Service ceiling: 8800m (29,000ft)

Dimensions: span 11.38m (37ft 4in); length 9.66m (31ft 8in); height 3.76m (12ft 4in)

Weight: 3760kg (8280lb) loaded

Armament: 6 x 12.7mm (0.5in) M2 Browning MGs in wings

▲ **Curtis Tomahawk Mk IIB**

No. 112 Squadron / North Africa, 1942

With a sturdy airframe and its Allison V-1710 engine optimized for low-altitude performance, the Tomahawk was better suited to the close support and tactical reconnaissance roles than the pure fighter task at higher altitudes.

▲ Grumman Wildcat Mk III

No. 805 Squadron / Fleet Air Arm, Western Desert, 1941

This aeroplane was flown over the Western Desert by Sub-Lieutenant Walsh.

Specifications

Crew: 1
Powerplant: 895kW (1200hp)
Maximum speed: 512km/h (318mph)
Range: 2012km (1250 miles)
Service ceiling: 10,365m (34,000ft)
Dimensions: span 11.58m (38ft); length 8.76m (28ft 9in); height 2.81m (9ft 5in)
Weight: 7952kg (3607lb) loaded
Armament: 6 x 12.7mm (0.5in) MGs in leading edge of wings

▲ Hawker Hurricane Mk IIB

No. 73 Squadron / Western Desert, 1942

Wearing a stylized version of the squadron's peacetime flash insignia, this Hurricane is in desert camouflage and has a Vokes filter under its nose.

Specifications

Crew: 1
Powerplant: 954kW (1280hp) Rolls-Royce Merlin XX liquid-cooled V-12
Maximum speed: 529km/h (329mph)
Range: 1480km (920 miles)
Service ceiling: 10,850m (35,600ft)
Dimensions: span 12.19m (40ft 0in); length 9.81m (32ft 2.25in); height 3.98m (13ft 1in)
Weight: about 3649kg (8044lb) loaded
Armament: 12 x 7.7mm (0.303in) Browning MGs and up to 454kg (1000lb) of bombs

▲ Hawker Hurricane Mk IIB

No. 73 Squadron / Western Desert, 1942

Wearing desert camouflage and a stylized version of the squadron's prewar marking on the side of the fuselage, this Hurricane Mk IIB has visible desert equipment in the form of the duct under the nose for the Vokes air filter.

Specifications

Crew: 1
Powerplant: 954kW (1280hp) Rolls-Royce Merlin XX liquid-cooled V-12
Maximum speed: 529km/h (329mph)
Range: 1480km (920 miles)
Service ceiling: 10,850m (35,600ft)
Dimensions: span 12.19m (40ft 0in); length 9.81m (32ft 2.25in); height 3.98m (13ft 1in)
Weight: about 3649kg (8044lb) loaded
Armament: 12 x 7.7mm (0.303in) Browning MGs and up to 454kg (1000lb) of bombs

▲ Hawker Hurricane Mk IIC

Royal Air Force / Western Desert, 1942

With its fixed forward-firing armament of four 20mm(0.79in) Hispano cannon, the Hurricane Mk IIC made an excellent close support and ground-attack fighter even though it was outmatched in the pure fighter role by German fighters such as the Messerschmitt Bf 109F.

Specifications

Crew: 1
Powerplant: 954kW (1280hp) Rolls-Royce Merlin XX liquid-cooled V-12
Maximum speed: 529km/h (329mph)
Range: 1480km (920 miles)
Service ceiling: 10,850m (35,600ft)
Dimensions: span 12.19m (40ft 0in); length 9.81m (32ft 2.25in); height 3.98m (13ft 1in)
Weight: about 3649kg (8044lb) loaded
Armament: 4 x 20mm (0.8in) Hispano cannon and up to 454kg (1000lb) of bombs

▲ Bristol Beaufighter Mk IC

No. 252 Squadron / RAF Coastal Command, Ecdu, Egypt, mid-1942

The aeroplane is a standard Coastal Command Beaufighter in the Middle East camouflage scheme of dark earth and mid-stone.

Specifications

Crew: 2
Powerplant: 2 x 1119kW (1500hp) Bristol Hercules XI air-cooled engine
Maximum speed: 492km/h (306mph)
Range: 2414km (1500 miles)
Service ceiling: 8810m (28,900ft)
Dimensions: span 17.63m (57ft 10in); length 12.60m (41ft 4in); height 4.82m (15 ft 10in)
Weight: 9435kg (21,100lb) loaded
Armament: 4 x 20mm (0.79in) cannon and 6 x 7.7mm (0.303in) MGs

▲ **Hurricanes over the Western Desert**

These are Hurricane Mk IIC tropicalized aircraft of No. 239 Squadron, with an undernose Vokes filter.

Battle for Malta

1940–43

The story of air warfare over the Mediterranean is, in effect, the story of the battle for Malta, which controls the east–west lines of communication that the British needed, and the north–south lines that the Italians and Germans needed.

THE AXIS SIEGE OF MALTA was of great import, and Malta become one of the most intensively bombed areas of World War II: some 3000 raids took place during the two years of the siege, and the civilian losses came to a total of some 1493 dead and 3674 wounded. Between June 1940 and December 1942, British and Commonwealth fighters claimed the destruction of some 863 Axis aircraft (the actual losses were about 570 aircraft in action) for the loss of 289 Spitfires and Hurricanes in action, and some 844 aircraft lost to all causes in the air and on the ground. The *Luftwaffe* alone claimed some 446 Allied aircraft (of all types) shot down.

ORGANIZATION

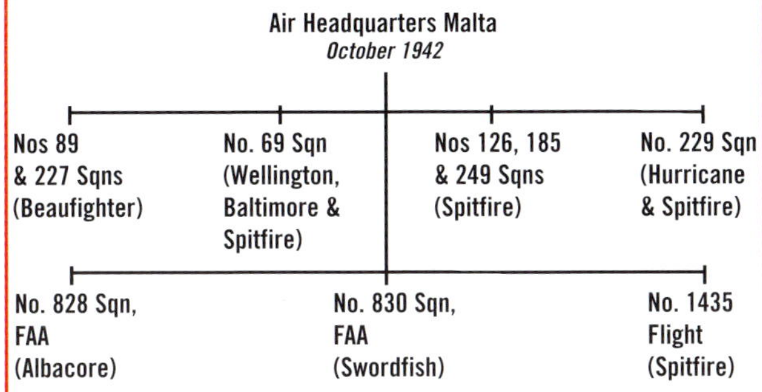

Faith, Hope and Charity

When Italy entered the war in June 1940, Malta was essentially undefended, as the British felt that the island could not be defended effectively. Thus there were only 4000 soldiers and a few obsolete biplanes on the island. The first Italian air attack came on 11 June, one day after Italy's declaration of war, and the failure of the initial attacks persuaded the British government that the island could indeed be held. It therefore sent reinforcements – a policy that was maintained for the rest of the war.

By the start of July, the Gladiators had been reinforced by Hawker Hurricanes to create No. 261 Squadron. More aircraft were delivered by aircraft carrier in August in the first of many such efforts. During the first five months of the siege, Malta's aircraft claimed some 37 Italian aircraft destroyed or damaged. In January 1941, the German X *Fliegerkorps* arrived in Sicily, and there was immediately a major increase in the bombing of Malta. The appearance in February of a *Staffel* of Bf 109E fighters of *Jagdgeschwader* 26 led to a rapid escalation in Hurricane losses: during the following four months, 7./JG 26 claimed 42 air victories without loss. In mid-1941, the British formed Nos 126 and 185 Squadrons, and the defence was bolstered by cannon-armed Hurricane Mk II fighters.

Beaufighter reinforcement

In April and May, the first Bristol Blenheim and Bristol Beaufighter attack warplanes arrived. The weight of the Axis attack declined later in the year as German units were diverted to the Eastern Front. But in December 1941 German forces turned their attention back to Malta.

All the necessities had to be brought in by sea, and resupply became very difficult and the island was almost cut off: 31 Allied ships were lost to bombing, and while the defence claimed some 191 aircraft shot down from June 1940 to December 1941, its own losses were 94 fighters.

▲ **Gloster Sea Gladiator Mk II**

RAF Air Defence of Malta, 1940

Flown by Sergeant Pilot Robertson, this Sea Gladiator Mk II was one of several assembled on Malta from spares left for the use of the Sea Gladiators embarked on the carrier HMS *Glorious* but found unused.

Specifications

Crew: 1
Powerplant: 298kW (400hp)
Maximum speed: 243km/h (151mph)
Endurance: 2 hours 45 minutes
Service ceiling: 7010m (23,000ft)
Dimensions: span 8.94m (29ft 4in); length 6.17m (20ft 3in); height 2.82m (9ft 3in)
Weight: 1189kg (2622lb) loaded
Armament: 2 x 7.7mm (.303in) Vickers MGs in upper part of forward fuselage

With the Hurricane now decidedly outclassed by the new Messerschmitt Bf 109Fs of JG 53 and Italian Macchi C.202s, the first Supermarine Spitfire Mk Vs flew into Malta in March 1942 , and the number of carrier deliveries increased through 1942 for the use of Nos 601 and 603 Squadrons.

By mid-1942, the Axis air strength against the island had reached its peak at about 520 German and 300 Italian aircraft. Throughout this period, Royal Navy submarines, RAF bombers and FAA torpedo aircraft operating from Malta continued to wreak havoc on Axis shipping, severely curtailing vital supplies and reinforcements to the German and Italian forces in North Africa. Even so, by this time the Axis powers thought that Malta had been effectively neutralized, and therefore that some of their strength could be diverted to other theatres.

George Cross

On 15 April 1942, King George VI awarded the island of Malta the George Cross, the highest civilian award for gallantry in the Commonwealth. In the first six months of 1942, there was only one 24-hour period without air raids. *Luftwaffe* records indicate that between 20 March and 28 April 1942, Malta was subjected to 11,819 sorties and 6557 tonnes (6452 tons) of bombs.

The British took advantage of the lull to fly in more fighters, but all other supplies remained in critically short supply. There was a new wave of attacks in October, but the Allied efforts in the Middle East were beginning to have their effect, and supplies were reaching Malta. As the Axis forces were progressively defeated in North Africa, the siege of Malta was lifted.

Tunisia

January–May 1943

Caught between Allied forces advancing from the west and east, and finding it almost impossible to get supplies of weapons and all other essentials, the Axis forces in North Africa were compressed into a lodgement in northern Tunisia and forced to surrender in May 1943.

Specifications

Crew: 2
Powerplant: 2 x 1119kW (1500hp) Bristol Hercules XI air-cooled engine
Maximum speed: 492km/h (306mph)
Range: 2414km (1500 miles)
Service ceiling: 8810m (28,900ft)
Dimensions: span 17.63m (57ft 10in); length 12.60m (41ft 4in); height 4.82m (15 ft 10in)
Weight: 9435kg (21,100lb) loaded
Armament: 4 x 20mm (0.79in) cannon and 6 x 7.7mm (0.303in) MGs in the wings, and 1 x 7.7mm (0.303in) Vicker rearward-firing MG in dorsal turret

▲ **Bristol Beaufighter Mk IC**

No. 272 Squadron / RAF, North African and Mediterranean theatre, 1942

The Beaufighter Mk IC was the variant of the Beaufighter Mk I heavy fighter optimized for the coastal role with improved radio and navigation equipment as well as more fuel. The type's cannon armament proved very effective against light coastal vessels and installations.

By the start of 1943, the advance of the British Eighth Army and Allied First Army, from east and west respectively, was placing the Axis forces in an increasingly difficult position. Without any significant Axis maritime lift capability to Tunisia, the Axis lodgement had been strengthened by air in November 1942, but Allied air power was now so rampant that any continuation of this process was impossible, as was air support for the Axis ground forces. However, although the Axis forces were trapped, the capable *Generalfeldmarschall* Erwin Rommel was nonetheless able to stall the Allies with a series of defensive operations, most notably with the battle of the Kasserine Pass, but the Axis forces were now outflanked, outgunned and outnumbered.

Under cover of increasingly powerful air forces, the Allies slowly squeezed the shattered Axis forces into northern Tunisia, where they capitulated on 13 May 1943, surrendering more than 275,000 men to become prisoners of war. It was a shattering defeat hastened by the skilful exercise of Allied air power.

Specifications

Crew: 1
Powerplant: 895kW (1200hp) Pratt & Whitney R-1830-86 14-cylinder engine
Maximum speed: 507km/h (315mph)
Range: 1851km (1150 miles)
Service ceiling: 10,900m (35,700ft)
Dimensions: span 11.58m 38ft); length 8.76m (28ft 8in); height 2.81m (9ft 3in)
Weight: 3607kg (7952lb) loaded
Armament: 6 x 12.5mm (0.5in) MGs in wings; 2 x 113kg (249lb) bombs

▲ Grumman Martlet Mk II

No. 888 Squadron / Fleet Air Arm / HMS* Formidable*, November 1942

The aeroplane is wearing US and British markings for use in *Torch*, the Allied amphibious landings in French northwest Africa in November 1942. Soon known as the Wildcat Mk II, this was a variant of the Martlet Mk I with folding rather than fixed wing panels and two more guns, and as such was effectively similar to the US Navy's F4F-4.

▲ Grumman Wildcat Mk IV

No. 888 Squadron / Fleet Air Arm / HMS* Formidable*, November 1942

This is the aeroplane flown by Squadron Leader Fleet for the Allied *Torch* landings in French northwest Africa, and wears a mix of British and US markings, the latter in an effort to persuade the Vichy French forces not to fire on any such aircraft.

Specifications

Crew: 1
Powerplant: 895kW (1200hp)
Maximum speed: 512km/h (318mph)
Range: 2012km (1250 miles)
Service ceiling: 10,365m (34,000ft)
Dimensions: span 11.58m (38ft); length 8.76m (28ft 9in); height 2.81m (9ft 5in)
Weight: 7952kg (3607lb) loaded
Armament: 6 x 12.7mm (0.5in) MGs in leading edge of wings

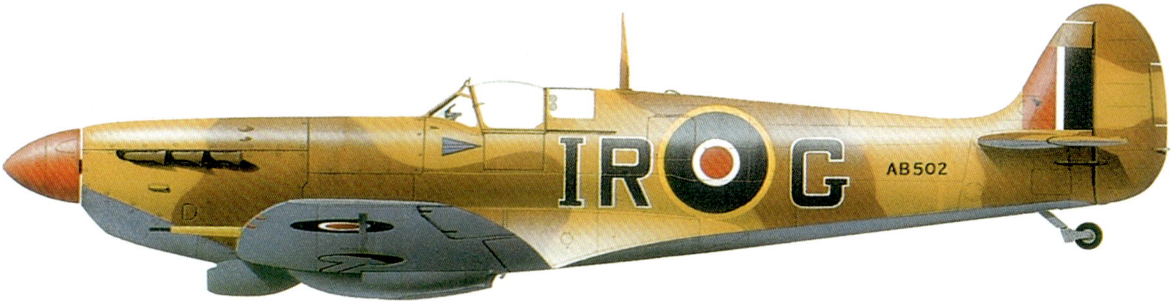

Specifications

Crew: 1
Powerplant: 1096kW (1470hp) Merlin 50 engine
Maximum speed: 594km/h (369mph)
Range: 1827km (1135 miles)
Service ceiling: 11,125m (36,500ft)
Dimensions: span 11.23m (36ft 10in); length 9.12m (29ft 11in); height 3.02m (9ft 11in)
Weight: 2911kg (6417lb) loaded
Armament: 4 x 7.7mm (0.303in) MGs and 2 x 20mm (0.8in) cannons in wings

▲ **Supermarine Spitfire Mk VB**

No. 224 Wing / Goubrine South, Tunisia, April 1943

This was the aeroplane in which Wing Commander Ian Gleed was shot down while trying to intercept and destroy German transport aircraft on 16 April 1943. Over the course of the war, 3923 Spitfire Mk VBs were built.

The Invasion of Sicily

July 1943

With their Axis foes driven from North Africa or captured, the Allies now turned their attention to a weakened Italy. The first step was clearly the short stride to the large island of Sicily, which would become a springboard for an Allied descent on the mainland.

THE ALLIED LANDINGS round the southern tip of Sicily began on 9 July 1943 and ended on 17 August with an Allied victory following a major Allied amphibious and airborne operation and subsequent advance. The assault formations were the US Seventh Army on the left and British Eighth Army on the right, while the Axis defence was made up of some 365,000 Italian and about 40,000 German troops.

Landings

The landings took place in a strong wind, which made them difficult, and two British and two US airborne drops were carried out just after midnight during the night of 9/10 July. The American paratroopers were mainly from the 505th Parachute Infantry Regiment of the 82nd Airborne Division, making their first combat drop. The strong winds caused aircraft to go off course and scattered them widely; the result was that around half the US paratroopers failed to reach their rallying points. British glider-landed troops fared little better, only one of 12 gliders landing on target, and many ditching at sea. Nevertheless, the scattered airborne troops maximized their opportunities, attacking patrols and creating confusion wherever possible.

Because of the adverse weather, many men were landed in the wrong place, wrong order and well behind time, but the British nonetheless took the port of Syracuse virtually unopposed. Only in the American centre was a substantial counterattack made, at exactly the point where the airborne force should have been. On 11 July, the US commander, Lieutenant-General George C. Patton, ordered his reserve parachute regiments to drop and reinforce the centre. Not every unit had been informed of the drop, and the 144 Douglas C-47 transports, which arrived shortly after an Axis air raid, were engaged by

▲ **Curtiss Kittyhawk Mk III**

No. 250 Squadron / RAF, southern Italy, autumn 1943

The Kittyhawk remained in useful service during the North African and Italian campaigns.

Specifications

Crew: 1
Powerplant: 895kW (1200hp) Allison piston engine
Maximum speed: 563km/h (350mph)
Range: 1738km (1080 miles)
Service ceiling: 9450m (31,000ft)
Dimensions: span 11.36m (37ft 4in); length 10.16m (33ft 4in); height 3.76m (12ft 4in)
Weight: 3511kg (7740lb) loaded
Armament: 4 x 12.7mm (0.5in) Browning MGs in wings

Allied warships: 33 were shot down and 37 damaged, resulting in 318 casualties to friendly fire. Even so, during the first two days progress was excellent; Vizzini in the west and Augusta in the east were taken. Then resistance in the British sector stiffened, and General Sir Bernard Montgomery persuaded the Allied army group commander to shift the inter-army boundaries so that the British could bypass resistance and retain the key role of capturing Messina.

After a week's fighting, Patton looked for a greater role for his army and opted to take Palermo, the capital. This spurred a coup against the Italian leader, Benito Mussolini, who was deposed. After the capture of Palermo, with the British still bogged down south of Messina, a two-pronged advance was ordered on this key port city. On 24 July, Montgomery suggested to Patton that his Seventh Army should take Messina, as it was better placed. The Seventh Army started its attack on what was now a German defence line at Troina, which held, and the Germans managed to keep the bulk of their forces beyond reach of capture and maintain their evacuation plans: men of the US 3rd Infantry Division entered Messina just after the last Axis troops left Sicily.

The Axis losses totalled 29,000, with 140,000 captured. The US Army lost 2237 killed and 6544 wounded and captured, while the British suffered 2721 dead and 10,122 wounded and captured. Allied air power was again dominant.

Italian Campaign

September 1943 – May 1945

The Allies landed on the mainland of Italy on 9 September 1943, and any hopes of swift victory were soon dashed. The Italians secured an armistice, but the Germans were masters of defensive warfare, and held out grimly in the face of overwhelming Allied air power.

On 9 September 1943, the US and British forces of the US Fifth Army landed against strong German opposition at Salerno on the 'shin' of the Italian 'leg' and elements of the British Eighth Army landed virtually unopposed at Taranto on the 'instep'. Italy had surrendered, but any Allied hopes that the Germans would pull out of Italy were unrealistic. The Eighth Army was able to make good progress for a time along the east coast, taking the port of Bari and the important airfields around Foggia. Despite the fact that it was not reinforced, the Germans came very close to defeating the landing at Salerno, which had been selected as it was within striking distance of the great port of Naples, the most northerly port that could be covered by Allied fighters operating from Sicilian bases.

ORGANIZATION

Desert Air Force
June 1944

No. 232 Bomber Wing	No. 239 Fighter-Bomber Wing	No. 285 Fighter-Bomber & Recce Wing	No. 3 Bomber Wing (SAAF)	No. 7 Fighter & Fighter-Bomber Wing (SAAF)

Natural defences

As the Allies advanced north, increasingly difficult mountain and river terrain checked all possibility of rapid advance and offered the German forces ideal opportunities for extended defence. Early in October 1943, Adolf Hitler was persuaded by his army group commander in southern Italy that the defence of Italy should be undertaken as far away from Germany as possible, and making optimum use of the natural defensive geography while at the same time denying the Allies the easy capture of a succession of airfields steadily closer to Germany.

Generalfeldmarschall Albert Kesselring was now given command of the whole of Italy, and he ordered the construction of a series of defence lines across Italy to the south of Rome.

Two lines, the Volturno Line and the Barbara Line, were used to delay the Allies and so yield the opportunity to prepare the most formidable defensive positions that formed the Winter Line. The Winter Line was the overall designation of the Gustav Line and two auxiliary defence lines, the Bernhardt and Adolf Hitler Lines, to the west of the Apennine mountains. The Winter Line proved a major obstacle to the Allies at the end of 1943, halting the advance of the Fifth Army on the western side of Italy. Although the Eighth Army broke through the Gustav Line in the east and took Ortona, truly dreadful weather at the very end of 1943 brought Allied efforts to a halt.

The Allies then focused on the western front where an attack through the Liri valley would open the way to Rome. Landings at Anzio behind the line were intended to destabilize the Germans' Gustav Line defence, but the anticipated early thrust inland to cut off the German defence did not happen and the Anzio force was held.

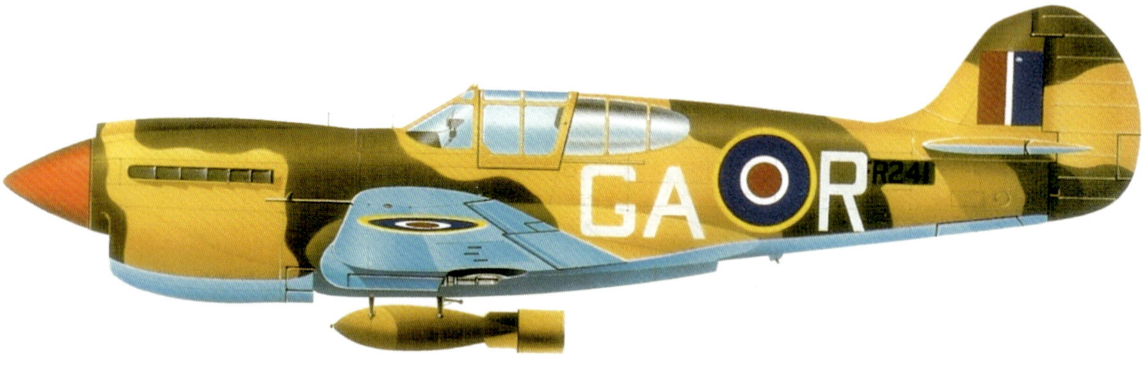

▲ Curtis Kittyhawk Mk IV

No. 112 Squadron / No. 239 Wing, Cutella, Italy, 1944

112 Squadron operated the Tomahawk IIB from July to December 1941 and at this time adopted the Sharkmouth markings for which it is now famous. The underfuselage hardpoint of the 'Kittybomber' could carry one 113kg (250lb) bomb, or a 227kg (500lb) bomb as shown here, or a 454kg (1000 kg) bomb.

Specifications

Crew: 1
Powerplant: 895kW (1200hp) Allison piston engine
Maximum speed: 563km/h (350mph)
Range: 1738km (1080 miles)
Service ceiling: 9450m (31,000ft)
Dimensions: span 11.36m (37ft 4in); length 10.16m (33ft 4in); height 3.76m (12ft 4in)
Weight: 3511kg (7740lb) loaded
Armament: 4 x 12.7mm (0.5in) Browning MGs in wings

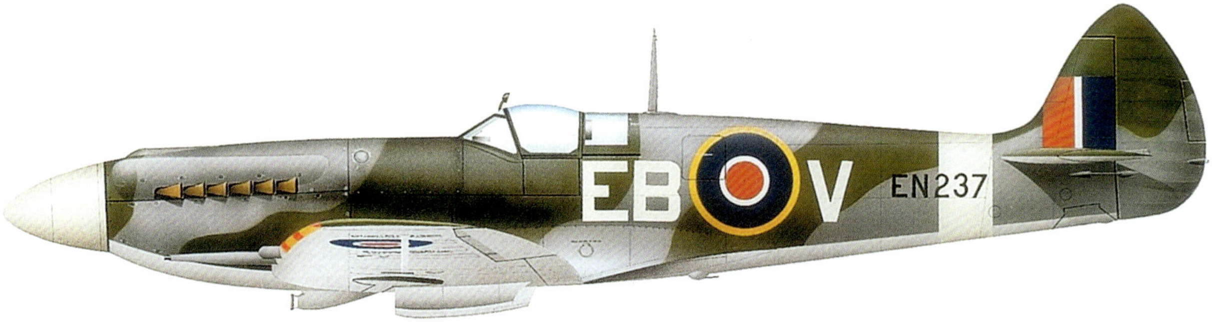

Specifications

Crew: 1
Powerplant: 1170kW (1565hp) 12-cylinder Rolls-Royce Griffon 2 engine
Maximum speed: 642km/h (410mph)
Range: 698km (435 miles) on internal fuel tanks
Service ceiling: 12,650m (41,500ft)
Dimensions: span 11.23m (36ft 10in); length 9.47m (31ft 1in); height 3.86m (12ft 8in)
Weight: 3343kg (7370lb) loaded
Armament: 4 x 7.7mm (0.303in) MGs and 2 x 20mm (0.8in) cannons

▲ **Supermarine Spitfire Mk XII**

No. 41 Squadron / Italy, 1944

Later-mark Supermarine Spitfire fighters were characterized by enlargement of the vertical tail surface, including a broader-cord rudder to maintain directional authority despite the larger nose with its Rolls-Royce Griffon engine.

▲ **North American Mustang Mk III**

No. 112 Squadron / RAF, Italy, 1945

This was the aeroplane of Flight Lieutenant Raymond V. Hearn, commander of the Squadron's 'B' Flight. Hearn was killed by German anti-aircraft fire on 18 February 1945 during his last scheduled sortie. The nose is painted with 112 Squadron's famous 'sharkmouth' markings.

Specifications

Crew: 1
Powerplant: 1081kW (1450hp) V-1650-3 engine
Maximum speed: 690km/h (430mph)
Range: 1215km (755 miles)
Service ceiling: 12,649m (41,500ft)
Dimensions: span 11.27m (37ft); length 9.84m (32ft 4in); height 4.15m (13ft 8in)
Weight: 4173kg (9200lb) loaded
Armament: 4 x 12.7mm (0.5in) MGs

It took four major offensives between January and May 1944 before the line was eventually broken by a combined assault of the Fifth and Eighth Armies concentrated along a 32km (20-mile) front between Monte Cassino and the west coast.

At the same time, the forces at Anzio broke out of their beachhead, but Fifth Army preferred to take Rome, on 4 June, rather than cut off sizeable German forces. From June to September, the Allies advanced beyond Rome, taking Florence and approaching the Gothic Line. This last major defensive line, just south of Bologna, was penetrated during the autumn campaign, but there was no decisive breakthrough until April 1945.

Then the German armies, shattered and virtually without fuel as a result of the domination of the Allied air forces, which also decimated all ground movement, finally collapsed.

RAF Fighter Command
1942–45

The years 1942 and 1943 were marked by battles of attrition with the *Luftwaffe*, but as German strength declined and Fighter Command started to introduce new fighters as well as advanced versions of its current fighters, the British gained a clear ascendancy over the *Luftwaffe*.

THE PROCESS OF WEARING down the *Luftwaffe* over France, which Fighter Command had started in 1941, continued in various forms during 1942–43. However, the expansion of the war effort in other theatres resulted in the outward transfer of many experienced pilots and squadrons. Thus the units left to No. 11 Group found it increasingly difficult to gain superiority over the small but very capable

Specifications

Crew: 1
Powerplant: 858kW (1150hp) Allison V-1710-39 V12 engine
Maximum speed: 690km/h (430mph)
Range: 1215km (755 miles)
Service ceiling: 12,649m (41,500ft)
Dimensions: span 11.27m (37ft); length 9.84m (32ft 4in); height 4.15m (13ft 8in)
Weight: 4173kg (9200lb) loaded
Armament: 4 x 12.7mm (0.5in) and 4 x 7.62mm (0.3in) Browning MGs

▲ **North American Mustang Mk I**

No. 414 (Canadian) Squadron / RAF Army Cooperation Command, Middle Wallop, 1942

The RAF received 620 examples of the Mustang Mk I, which replaced the Curtiss Tomahawk in service with Army Cooperation squadrons for the reconnaissance-fighter role. There were also 150 cannon-armed Mustang Mk IAs and 50 long-range Mustang Mk II aircraft.

Specifications

Crew: 1
Powerplant: 1102kW (1478hp) Rolls-Royce Merlin 45 engine
Maximum speed: 594km/h (369mph)
Range: 1827km (1135 miles)
Service ceiling: 11,125m (36,500ft)
Dimensions: span 11.23m (36ft 10in); length 9.12m (29ft 11in); height 3.02m (9ft 11in)
Weight: 2911kg (6417lb) loaded
Armament: 4 x 7.7mm (0.303in) MGs and 2 x 20mm (0.8in) cannons

▲ **Supermarine Spitfire Mk VC**

No. 91 Squadron / RAF Fighter Command, Hawkinge, 1942

The Spitfire Mk VC was the development of the Spitfire Mk V with the 'universal' wing able to carry machine-gun or cannon/machine-gun armament. The wing could also have its tips removed for a greater roll rate at low altitude, or extended for greater ceiling. This has the 'B' type cannon/machine-gun armament.

ORGANIZATION

Fighter Command
March 1943

- No. 9 Group
- No. 10 Group
- No. 11 Group
- No. 12 Group
- No. 13 Group
- No. 14 Group
- No. 60 Signals Group
- No. 81 Fighter Training Group

fighter units that the Germans left in France. The squadrons were additionally committed to exhausting defensive patrols against small units of Focke-Wulf Fw 190 fighter-bombers flying low-level 'hit and run' raids against ports and towns along the south coast of England. Yet the new Hawker Typhoon fighter proved itself able to catch these raiders.

Dieppe raid

The most notable offensive battle took place over the 'reconnaissance in force' against Dieppe in August 1942. The *Luftwaffe* and RAF fought it out over the failed assault, and though Fighter Command managed to prevent the *Luftwaffe* from attacking the assault shipping (its primary objective) the British success proved illusory. Despite the claims of the day that more German than British aircraft had been shot down, later analysis revealed exactly the opposite.

In 1943, the most notable event was the division of Fighter Command into the Air Defence of Great Britain (ADGB) and the 2nd Tactical Air Force (TAF). As its designation indicates, the ADGB was entrusted with the task of defending the UK from air attack, while the 2nd TAF was tasked with the support of ground forces after the eventual invasion of Europe.

The year 1944 saw the ADGB's greatest effort. *Overlord*, the Allied invasion of northern France, was launched on 6 June 1944. British fighters wove over the battle area and, together with their US brothers in arms, completely suppressed the efforts of the miniscule German opposition.

They also directly supported ground forces by strafing enemy positions and transport. Later in the year, the final major trial of what had become Fighter Command once again in October 1944, occurred with the defeat of the V-1 flying bomb campaign. In World War II, Fighter Command lost 3690 men killed, 1215 wounded and 601 taken prisoner, as well as 4790 aircraft lost.

Specifications

Crew: 1
Powerplant: 1170kW (1565hp) 12-cylinder Rolls-Royce Merlin 61 engine
Maximum speed: 642km/h (410mph)
Range: 698km (435 miles) on internal fuel tanks
Service ceiling: 12,650m (41,500ft)
Dimensions: span 11.23m (36ft 10in); length 9.47m (31ft 1in); height 3.86m (12ft 8in)
Weight: 3343kg (7370lb) loaded
Armament: 4 x 7.7mm (0.303in) MGs and 2 x 20mm (0.8in) cannons

▲ Supermarine Spitfire Mk IX

Special Service Flight, Northolt, 1942

Although developed as an interim type with the airframe of the Spitfire Mk V and the Rolls-Royce Merlin 60 series engine, the Spitfire Mk IX was produced in larger numbers than any other Spitfire variant.

Specifications

Crew: 1
Powerplant: 1170kW (1565hp) 12-cylinder Rolls-Royce Merlin 61 engine
Maximum speed: 642km/h (410mph)
Range: 698km (435 miles) on internal tanks
Service ceiling: 12,650m (41,500ft)
Dimensions: span 11.23m (36ft 10in); length 9.47m (31ft 1in); height 3.86m (12ft 8in)
Weight: 3343kg (7370lb) loaded
Armament: 4 x 7.7mm (0.303in) MGs and 2 x 20mm (0.8in) cannons

▲ Supermarine Spitfire Mk IXE

Kenley Wing / RAF Fighter Command, Kenley, 1943–44

This was the personal aeroplane of Wing Commander J. E. 'Johnnie' Johnson, commanding the Kenley Wing and later No. 127 (Canadian) Wing at Kenley, and as such sporting his initials rather than a squadron coding.

Night-fighters
1942–45

The Air Defence of Great Britain was one part of Fighter Command after this had been divided to allow the creation of the 2nd Tactical Air Force. Although only short-lived, the ADGB has its place in history as the air formation that helped to defeat the V-1 flying bomb campaign.

THE FIRST INCARNATION of the Air Defence of Great Britain (ADGB) happened in 1925, as the command of this name was created for the supervision of home defences, which at the time

Specifications

Crew: 2
Powerplant: 2 x 918kW (1230hp) Rolls-Royce Merlin XX engines
Maximum speed: 612km/h (380mph)
Range: 1963km (1220 miles)
Service ceiling: 9449m (31,000ft)
Dimensions: span 16.51m (54ft 2in); length 12.43m (40ft 10in); height 4.65m (15ft 3in)
Weight: 5942kg (13,100lb) loaded
Radar: AI.Mk IV interceptor radar
Armament: 4 x 20mm (0.8in) Hispano cannon and 4 x 7.7mm (0.303in) Browning MGs

▲ de Havilland Mosquito NF.Mk II

No. 157 Squadron / RAF Fighter Command, Castle Camps, mid-1942

The first fighter Mosquito introduced into service was the NF. Mk II in mid-1942. The Mosquito NF. Mk II brought an altogether better capability to the RAF's night-fighter arm, for it offered a much higher level of performance than the Beaufighter.

ORGANIZATION

Circus No. 258
3 February 1943

- **Bomber Force** (12 Venturas of No. 21 Squadron)
- **Close Escort Force** (3 x Spitfire squadrons)
- **1st Fighter Echelon** (3 x squadrons of the Biggin Hill Wing)
- **2nd Fighter Echelon** (4 x Spitfire squadrons of the Kenley Wing)

included bombers as well as fighters. This first ADGB was split into two areas controlling regular squadrons. These were the Wessex Bombing Area and the Fighting Area. In 1936, the ADGB was disestablished, whereupon the erstwhile Bombing Area became Bomber Command, and the Fighting Area became Fighter Command.

The second incarnation of the Air Defence of Great Britain designation came when it was adopted for what was left of Fighter Command after the 2nd Tactical Air Force (TAF) had been hived off Fighter Command in 1943. The new ADGB was tasked with the air defence of Britain.

It was nonetheless nothing more than a small Fighter Command, and this fact was recognized in 1944 when the command reverted to the Fighter Command designation.

▲ **Bristol Beaufighter Mk II**

No. 307 (Polish) Squadron / Exeter, April 1943

The red-and-white checkerboard marking, derived from the Polish flag, was a standard adornment of the aircraft flown by Polish-manned squadrons.

Specifications

Crew: 2
Powerplant: 2 x 954kW (1280hp) Rolls-Royce Merlin XX engines
Maximum speed: 492km/h (306mph)
Range: 2414km (1500 miles)
Service ceiling: 8810m (28,900ft)
Dimensions: span 17.63m (57ft 10in); length 12.60m (41ft 4in); height 4.82m (15 ft 10in)
Weight: 9435kg (21,100lb) loaded
Armament: 4 x 20mm (0.79in) cannon and 6 x 7.7mm (0.303in) MGs in the wings

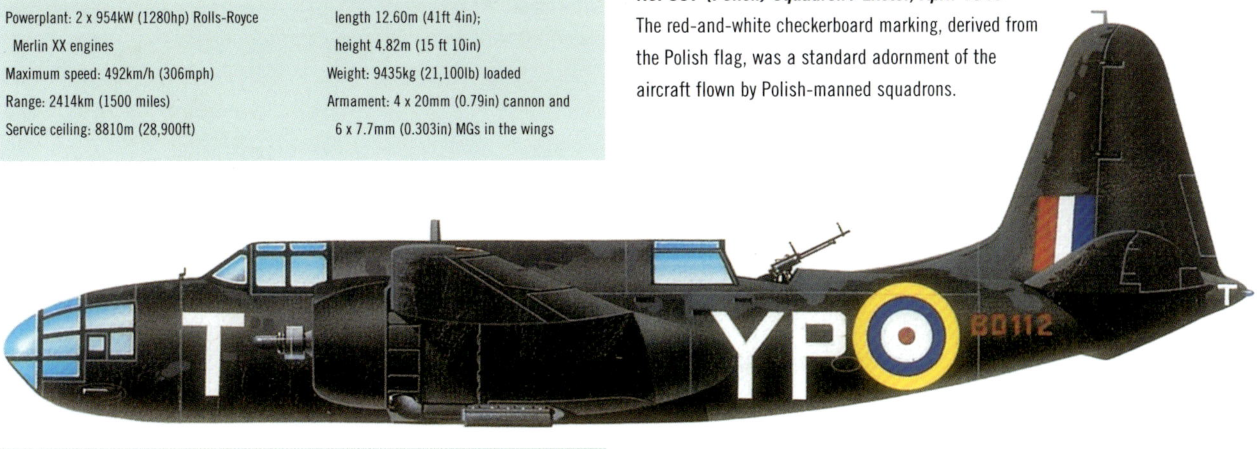

▲ **Douglas Havoc Mk I**

No. 23 Squadron / RAF Fighter Command, Ford, 1943

The Havoc fighter development of the DB-7 bomber made a moderately useful interim night-fighter, but was too large an aeroplane for the role.

Specifications

Crew: 3
Powerplant: 1193kW (1600hp) Wright R-2600-23 radial piston engine
Maximum speed: 510km/h (317mph)
Range: 1521km (945 miles)
Service ceiling: 7225m (23,700ft)
Dimensions: span 18.69m (61ft 4in); length 14.63m (47ft 11in); height 5.36m (17ft 7in)
Weight: 10,964kg (24,127lb) loaded
Armament: 6 x 12.7mm (0.5in) forward-firing Browning M2 MGs; 2 x 2.7mm (0.5in) MGs in power-operated dorsal turret; 2 x 2.7mm (0.5in) rearward firing MGs in ventral tunnel

Flying bomb interceptors
1944

The first of Germany's 'vengeance' weapons was the V-1 flying bomb, a primitive cruise missile guided by an autopilot and powered by a pulsejet. The V-1 was comparatively fast at low level and inaccurate, but it carried a large warhead and was a real menace to civilian morale.

THE CAMPAIGN AGAINST the V-1 involved balloon barrages, belts of anti-aircraft guns and fighters. The latter recorded its first interception on 14/15 June 1944. When the V-1 campaign began in mid-June of 1944, there were fewer than 30 examples of the Hawker Tempest, the only fighter with the low-altitude speed needed to catch a V-1, and these were allocated to No. 150 Wing.

Tempest wing

The Tempest wing was increased to more than 100 aircraft by September. P-51s and Griffon-engined Spitfire Mk XIVs were also specially tuned to give them the speed needed, and at night de Havilland Mosquitoes were used. There was no need for radar since the V-1's exhaust plume was visible from far away. Daylight V-1 chases were chaotic and often unsuccessful until a special defence zone was declared between London and the coast, in which only the fastest fighters were permitted to operate.

Between June and 5 September 1944, the handful of No. 150 Wing Tempests shot down 638 flying bombs, No. 3 Squadron claiming 305. Next most successful were the Mosquito (428), Spitfire Mk XIV (303) and Mustang (232).

Meteor speed

Even though it was not fully operational, the jet-powered Gloster Meteor was rushed into service with No. 616 Squadron RAF to fight the V-1s. It had ample speed, but its cannon were prone to jamming. By September 1944, the V-1 threat to England was removed when all its launch sites were overrun by the advancing Allied armies. Some 4261 V-1s had been destroyed by fighters, anti-aircraft fire and barrage balloons.

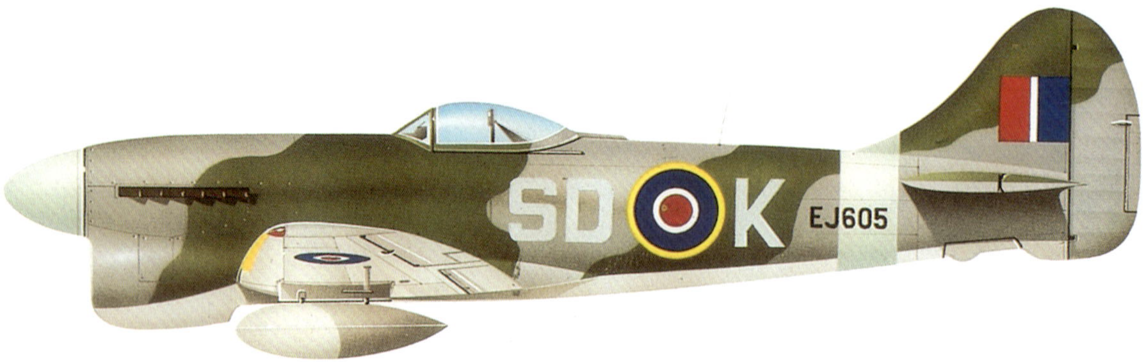

▲ **Hawker Tempest Mk V (Series 2)**

No. 501 Squadron / RAF, 1944–45

A conceptual development of the Typhoon with a thinner wing, among other changes, the Tempest was an altogether superior fighter. Delivered in August 1944, the Tempests of 501 Squadron specialized in defensive operations against the V-1 flying bombs.

Specifications

Crew: 2
Powerplant: 1626kW (2180hp) Napier Sabre IIA H-type piston engine
Maximum speed: 686km/h (426mph)
Range: 2092km (1300 miles)
Service ceiling: 10,975m (36,000ft)
Dimensions: span 12.50m (41ft); length 10.26m (33ft 8in); height 4.90m (16ft 1in)
Weight: 6142kg (13,540lb) loaded
Armament: 4 x 20mm (0.8in) Hispano cannon in wings, plus up to 907kg (2000lb) disposable stores consisting of either 2 x bombs or 8 rockets for ground attack role

Specifications

Crew: 1
Powerplant: 1081kW (1450hp) V-1650-3 engine
Maximum speed: 690km/h (430mph)
Range: 1215km (755 miles)
Service ceiling: 12,649m (41,500ft)
Dimensions: span 11.27m (37ft); length 9.84m (32ft 4in); height 4.15m (13ft 8in)
Weight: 4173kg (9200lb) loaded
Armament: 4 x 12.7mm (0.5in) MGs

▲ North American Mustang Mk IIIB

No. 316 (Polish) Squadron / Coltishall, June 1944

The Mustang Mk III was the British counterpart of the USAAF's P-51B and was employed for a time as flying bomb interceptor.

Specifications

Crew: 1
Powerplant: 1081kW (1450hp) V-1650-3 engine
Maximum speed: 690km/h (430mph)
Range: 1215km (755 miles)
Service ceiling: 12,649m (41,500ft)
Dimensions: span 11.27m (37ft); length 9.84m (32ft 4in); height 4.15m (13ft 8in)
Weight: 4173kg (9200lb) loaded
Armament: 4 x 12.7mm (0.5in) MGs

▲ North American Mustang Mk III

No. 306 'Torunski' (Polish) Squadron / UK, 1944

This aeroplane has the British 'Malcolm hood' in place of the original framed canopy, much improving the pilot's fields of vision.

Specifications

Crew: 1
Powerplant: 2 x 7.56kN (1700lb st) Rolls-Royce W.2B/23C Welland turbojet engines
Maximum speed: 668km/h (415mph)
Service ceiling: 12,190m (40,000ft)
Dimensions: span 13.11m (43ft); length 12.57m (41ft 3in); height 3.96m (13ft)
Weight: 6257kg (13,795lb) maximum take-off
Armament: 4 x 20mm (0.8in) Hispano cannon fixed forward-firing in nose

▲ Gloster Meteor Mk I

No. 616 Squadron / RAF, 1945

Although it offered better performance than the piston-engined fighters of the day, the Meteor was little more than a turbojet-powered development of an airframe that was otherwise typical of piston-engined thinking.

RAF 2nd Tactical Air Force
1943–45

The 2nd Tactical Air Force was created using the tactics developed in North Africa by the Desert Air Force and then translated to the Italian campaign of 1943. It was designed to provide the British and Canadian forces with superb air support in the northwest Europe campaign.

THE 2ND TACTICAL AIR FORCE (TAF) was one of the three tactical air forces that were established within the Royal Air Force during and after World War II.

The 2nd TAF comprised squadrons and personnel from the air forces of the British Commonwealth and the air units of various governments in exile, as well as from the RAF itself.

ORGANIZATION

2nd Tactical Air Force (Air Marshal Sir Arthur Coningham)
June 1944

No. 2 Group (Air Vice Marshal B. E. Embry)	No. 83 Group (Air Vice Marshal H. Broadhurst)	No. 84 Group (Air Vice Marshal L. O. Brown)	No. 85 Group (Air Vice Marshal J. B. Cole-Hamilton)

2nd TAF established

The 2nd TAF was established during June 1943, even before the DAF (1st TAF) had moved from North Africa to Italy to perfect its tactics under European mainland conditions. It was part of the Allied preparations already advanced for the invasion of northwest Europe, an operation planned for the summer of 1944.

The 2nd TAF drew units from many elements of the RAF commands in the UK in order to build up a force able to provide a high level of support for the British Army in the field.

Bomber Command provided its No. 2 Group, which was equipped with light bombers and fighter-bombers, and Fighter Command was divided into two portions, namely the Air Defence of Great Britain, which retained fighter units for the air defence of the UK, and Nos 83 and 84 Groups for allocation to the 2nd TAF.

▲ **Supermarine Spitfire Mk IXE**

No. 144 (Canadian) Wing / Kenley, March 1943

This was the personal aeroplane of Wing Commander J. E. 'Johnnie' Johnson.

Specifications

Crew: 1
Powerplant: 1170kW (1565hp) 12-cylinder Rolls-Royce Merlin 61 engine
Maximum speed: 642km/h (410mph)
Range: 698km (435 miles) on internal fuel tanks
Service ceiling: 12,650m (41,500ft)
Dimensions: span 11.23m (36ft 10in); length 9.47m (31ft 1in); height 3.86m (12ft 8in)
Weight: 3343kg (7370lb) loaded
Armament: 4 x 7.7mm (0.303in) MGs and 2 x 20mm (0.8in) cannons

The new force's first commander was Air Marshal Sir John d'Albiac. But on 21 January 1944, d'Albiac was followed by the commander most usually associated with the 2nd TAF, namely Air Marshal Sir Arthur Coningham.

This officer possessed vast experience of the type of air operations needed for the effective support of fast-moving land warfare, for he had commanded the DAF in North Africa and Italy, and now perfected the 2nd TAF to meet any and all the demands that would be placed on it. One of the primary tactics was the 'cab rank' system of on-call close support developed in North Africa and Italy.

By mid-1944, the *Luftwaffe* was nothing like the force it once had been, so air opposition to the 2nd TAF was limited. The force could therefore concentrate on the support of the British and Canadian forces on the left flank of the Allied invasion of France and subsequent advance east toward Germany.

WINGS OF NO. 83 GROUP, 2ND TAF (JUNE 1944)

Wing	Squadrons	Aircraft
No. 39, RCAF	168 & 414 430	Mustang Spitfire
No. 121	174, 175 & 245	Typhoon
No. 122	19, 65 & 122	Typhoon
No. 122	19, 65 & 122	Mustang
No. 124	181, 182 & 247	Typhoon
No. 125	132, 453 & 602	Spitfire
No. 126, RCAF	401, 411 & 412	Spitfire
No. 127, RCAF	403, 416 & 421	Spitfire
No. 129, RCAF	184	Typhoon
No. 143, RCAF	438, 439 & 440	Typhoon
No. 144, RCAF	441, 442 & 443	Spitfire
Air observation post	652, 653, 658, 659 & 662	Auster

Bodenplatte

One exception to this general situation came with *Bodenplatte*, the *Luftwaffe*'s last major effort launched on 1 January 1945, when the 2nd TAF suffered serious losses on the ground. But the *Luftwaffe's* standards of training were now so low that many aircraft were shot down by ground fire or Allied fighters, and others ran out of fuel and crashed because of their pilots' navigational errors.

Specifications

Crew: 1

Powerplant: 1685kW (2260hp) Napier Sabre II liquid cooled H-24 in-line piston engine

Maximum speed: 650km/h (405mph)

Range: 980km (610 miles)

Service ceiling: 10,400m (34,000ft)

Dimensions: span 12.67m (41ft 7in); length 9.73m (31ft 11in); height 4.66m (15ft 4in)

Weight: 5170kg (11,400lb) loaded

Armament: 4 x 20mm (0.8in) Hispano-Suiza HS.404 cannons, 2 x 454 kg (1000lb) bombs

▲ **Hawker Typhoon Mk IB**

No. 247 Squadron / No. 124 Wing / No. 83 Group / 2nd TAF, 1944

Although the Typhoon is best known in the attack role, with an underwing load of eight 76mm (3in) unguided rockets, it also operated in the close support role with two bombs of up to 454kg (1000lb) size under the wing.

▲ **Hawker Typhoon Mk IB**

No. 198 Squadron / No. 123 Wing / No. 84 Group / 2nd TAF

Depicted in the time before the *Overlord* invasion of Normandy, which it supported, this Typhoon Mk IB lacks 'invasion stripe' markings.

Specifications

Crew: 1

Powerplant: 1685kW (2260hp) Napier Sabre II liquid cooled H-24 in-line piston engine

Maximum speed: 650km/h (405mph)

Range: 980km (610 miles)

Service ceiling: 10,400m (34,000ft)

Dimensions: span 12.67m (41ft 7in); length 9.73m (31ft 11in); height 4.66m (15ft 4in)

Weight: 5170kg (11,400lb) loaded

Armament: 4 x 20mm (0.8in) Hispano-Suiza HS.404 cannons, 2 x 454 kg (1000lb) bombs

Normandy landings

June–August 1944

The period between the landing in Normandy (6 June) and the German escape from the Falaise pocket (20 August) saw the use of great Allied airpower, but the development of this capability was hindered by operations from the UK until airfields had been captured or created in France.

THE ALLIED INVASION and reconquest of northwest Europe started on 6 June 1944 with the launch of *Overlord*, the Allied 21st Army Group's amphibious landings in Normandy under cover of Air Chief Marshal Sir Trafford Leigh-Mallory's Allied Expeditionary Air Force (AEAF). The primary British and Commonwealth component of the AEAF was Air Marshal Sir Arthur Coningham's 2nd TAF operating from bases in southern England.

ORGANIZATION

No. 2 Group, 2nd TAF (Air Vice Marshal B. E. Embry)

June 1944

- No. 137 Wing (three squadrons)
- No. 138 Wing (three squadrons)
- No. 139 Wing (three squadrons)
- No. 140 Wing (three squadrons)

The Germans were taken completely by surprise, and in any case had only small air forces in France. The Allies, therefore, had matters completely their own way on 6 June as the light bombers of the 2nd TAF aided the heavy bombers of Bomber Command in attacking beach defences and installations, and then moved their attentions inland to destroy the Germans' last surviving lines of communication as the assault forces progressed south.

Here the light bombers collaborated with the 2nd TAF's attack aircraft under a cover of late-generation fighters, while low-level tactical reconnaissance planes swept ahead of the advancing ground forces in search of German defensive strongpoints and the possible arrival of German reinforcements, especially of armour. During that day, Allied air forces flew 14,674 sorties and lost only 113 aircraft, some of the losses being unfortunate victims of 'friendly fire'. The *Luftwaffe* was able to respond with a mere 319 sorties, which achieved only negligible results.

The nature of the campaign that was now under way was reflected on the following day, when the warplanes of Nos 83 and 84 Groups attacked a German Panzer division moving toward the front, destroying 90 trucks, 40 fuel tankers, five tanks and 84 other armoured vehicles, including assault guns and self-propelled artillery. The Allies gained the use of their first airstrip in Normandy on 10 June, and in the British and Commonwealth portion of the lodgement 31 more such strips were later developed, allowing the transfer of squadrons from the UK and thus longer endurance over the battlefield.

Luftwaffe response

The *Luftwaffe* responded as and when it could in the face of the essentially total air superiority of the Allies, who could thus use their tactical air power as and when they weather permitted. Even so, the Germans performed wonders in getting men, tanks, artillery and essential supplies to the front, and so seriously delayed the 21st Army Group. The German front finally began to unravel on 24 July as the Americans broke out in the west and then swung round to the east. The Germans were now threatened with a major encirclement in a pocket near Falaise, but managed to extricate many of their men before the neck of the pocket could be closed.

Even so, it was a major disaster for the Germans at Falaise, who lost most of their armour, artillery and motor transport. The pocket held 19 German divisions, including nine Panzer divisions. Before they could escape, the Allied tactical squadrons, spearheaded by Hawker Typhoon fighter-bomber units, effectively destroyed the 5. and 7. *Panzerarmees* over a three-day period (17–19 July), destroying or forcing the abandonment of all but about 120 of 2300 German armoured vehicles.

The Germans tried to respond in the air, but the *Luftwaffe* was only a shadow of its former self and by mid-August *Luftflotte 3* had been reduced to just 75 aircraft.

▲ Fighter bomber

Armed with eight rocket projectiles, this Typhoon Mk IB carries a typical Typhoon warload for low-level missions.

▲ Hawker Typhoon IB

No. 486 Squadron / Air Defence of Great Britain, spring 1944

Not all Hawker Typhoon fighter-bomber units were allocated to the 2nd TAF, this squadron remaining on the strength of Fighter Command (later the ADGB) at Tangmere on the coast of Sussex in southern England.

Specifications

Crew: 1

Powerplant: 1685kW (2260hp)

Maximum speed: 650km/h (405mph)

Range: 980km (610 miles)

Service ceiling: 10,400m (34,000ft)

Dimensions: span 12.67m (41ft 7in); length 9.73m (31ft 11in); height 4.66m (15ft 4in)

Weight: 5170kg (11,400lb) loaded

Armament: 4 x 20mm (0.8in) Hispano-Suiza HS.404 cannons, 8 x 27kg (60lb) rockets for ground attack purposes

Specifications

Crew: 1
Powerplant: 1685kW (2260hp) Napier Sabre II piston engine
Maximum speed: 650km/h (405mph)
Range: 980km (610 miles)
Service ceiling: 10,400m (34,000ft)
Dimensions: span 12.67m (41ft 7in); length 9.73m (31ft 11in); height 4.66m (15ft 4in)
Weight: 5170kg (11,400lb) loaded
Armament: 4 x 20mm (0.8in) Hispano-Suiza HS.404 cannons, 8 x 27kg (60lb) rockets for ground attack purposes

▲ Hawker Typhoon Mk IB

No. 193 Squadron / No. 146 Wing / No. 84 Group / 2nd TAF

Seen here wearing northern European day-fighter camouflage and black-and-white recognition strips for the invasion of Europe, the Typhoon Mk IB proved itself a decisive weapon in the northwestern Europe campaign as an attack fighter well able to deal with German armoured vehicles and artillery.

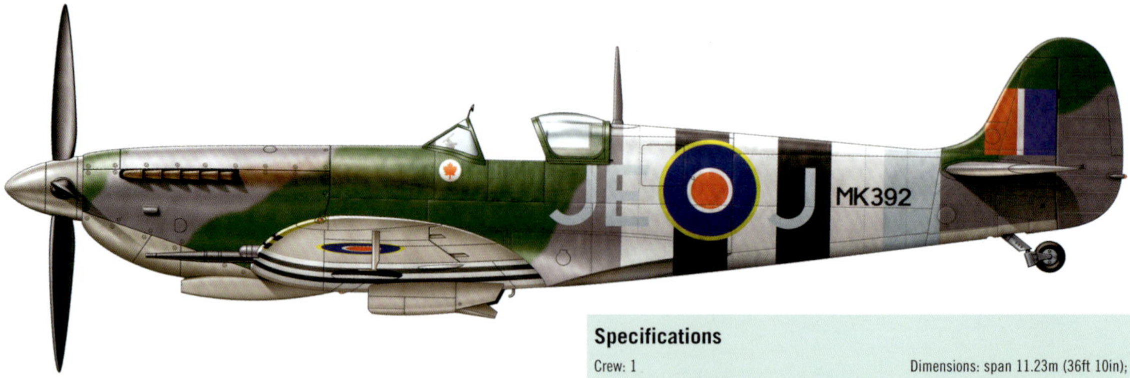

▲ Supermarine Spitfire Mk IX

No. 127 (Canadian) Wing / northwest Europe, autumn 1944

This was another of Wing Commander J. E. 'Johnnie' Johnson's personal aircraft, from the time he commanded another Canadian wing in the northwest European campaign from bases in France.

Specifications

Crew: 1
Powerplant: 1170kW (1565hp) 12-cylinder Rolls-Royce Merlin 61 engine
Maximum speed: 642km/h (410mph)
Range: 698km (435 miles) on internal fuel tanks
Service ceiling: 12,650m (41,500ft)
Dimensions: span 11.23m (36ft 10in); length 9.47m (31ft 1in); height 3.86m (12ft 8in)
Weight: 3343kg (7370lb) loaded
Armament: 4 x 7.7mm (0.303in) MGs and 2 x 20mm (0.8in) cannons

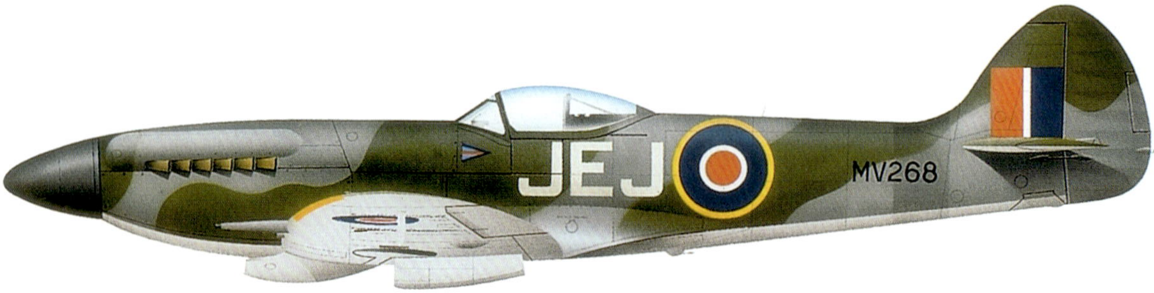

▲ Supermarine Spitfire Mk XIV

Northwest Europe, autumn 1944

The Mk XIV was used by the 2nd Tactical Air Force as their main high-altitude air superiority fighter in northern Europe. In total, 957 Mk XIVs were built, over 400 of which were FR Mk XIVs. This fighter also bears the personal markings of Wing Commander J. E. 'Johnnie' Johnson.

Specifications

Crew: 1
Powerplant: 1528kW (2050hp) Griffon 65 series engine
Maximum speed: 720km/h (448mph)
Range: 965km (600 miles) using additional rear fuel tank
Service ceiling: 13,560m (44,500ft)
Dimensions: span 11.23m (36ft 10in); length 9.96m (32ft 8in); height 3.86m (12ft 8in)
Weight: 3343kg (7370lb) loaded
Armament: 4 x 7.7mm (0.303in) MGs and 2 x 20mm (0.8in) cannons

Specifications

Crew: 1	Dimensions: span 11.23m (36ft 10in); length 9.47m (31ft 1in); height 3.86m (12ft 8in)
Powerplant: 1170kW (1565hp) 12-cylinder Rolls-Royce Merlin 61 engine	Weight: 3343kg (7370lb) loaded
Maximum speed: 642km/h (410mph)	Armament: N/A
Range: 698km (435 miles) on internal fuel tanks	
Service ceiling: 12,650m (41,500ft)	

▲ Supermarine Spitfire PR.Mk IX

Photographic reconnaissance squadron / RAF, 1944

Though carrying no armament, the Spitfire PR.Mk IX was very important to the Allied war effort – its ability to fly fast and high over long ranges made it an excellent photo-reconnaissance aeroplane.

▼ Supermarine Spitfire Mk VII

RAF Fighter Command, summer 1944

The Spitfire Mk VII was optimized for the high-altitude role with the Merlin 61 engine, a longer-span wing with more pointed tips and a retractable tailwheel. Production amounted to 471 aircraft, which served with 11 Squadrons based mostly in the UK.

Specifications

Crew: 1	Dimensions: span 11.23m (36ft 10in); length 9.47m (31ft 1in); height 3.86m (12ft 8in)
Powerplant: 1170kW (1565hp) 12-cylinder Rolls-Royce Merlin 61 engine	Weight: 3343kg (7370lb) loaded
Maximum speed: 642km/h (410mph)	Armament: 4 x 7.7mm (0.303in) MGs and 2 x 20mm (0.8in) cannons
Range: 698km (435 miles) on internal fuel tanks	
Service ceiling: 12,650m (41,500ft)	

▶ Spitfire Mk IXC

This is part of No. 306 *Torunski* Squadron, a Polish-manned unit based at Northolt in 1943–44 and engaged in daylight sweeps over Europe.

▲ **Hawker Tempest Mk V Series 2**

No. 274 Squadron, RAF / summer 1944

No. 274 Squadron converted from the Supermarine Spitfire Mk IX to the Tempest Mk V at Hornchurch during August 1944, and moved to Belgium in the following month. For the rest of World War II, the squadron flew armed reconnaissance missions, which led to many combats deep behind the frontline.

Specifications

Crew: 2
Powerplant: 1626kW (2180hp) Napier Sabre IIA H-type piston engine
Maximum speed: 686km/h (426mph)
Range: 2092km (1300 miles)
Service ceiling: 10,975m (36,000ft)
Dimensions: span 12.50m (41ft); length 10.26m (33ft 8in); height 4.90m (16ft 1in)
Weight: 6142kg (13,540lb) loaded
Armament: 4 x 20mm (0.8in) Hispano cannon in wings, plus up to 907kg (2000lb) disposable stores consisting of either 2 x bombs or 8 rockets for ground attack role

No. 2 Group

1944–45

No. 2 Group was formed on 20 March 1936, with five wings of ten squadrons. Four of the wings were allocated to the Advanced Air Striking Force (AASF), and a Bristol Blenheim of No. 2 Group made the first British operational sortie to cross the German frontier in World War II.

THE 2ND TAF's primary mission was to support the fighting troops on the ground. The fighters and fighter-bombers of the 83rd and 84th Groups were more suitable for what is now known as close-support – ground attack with rockets, cannon and bombs – in pinpoint and diving attacks against single targets such as tanks, trains, bridges and field fortifications. The agility of these light aircraft enabled them to make precision attacks in a way that the larger bombers, committed to horizontal bombing runs, could not.

In May 1943, No. 2 Group was detached from Bomber Command, now increasingly concerned with heavy night bombing, and allocated to the 2nd TAF. The group was very active in the northwest Europe campaign for tactical bombing, and in the closing months of the war was based on the continent. No. 2 Group started the war with 79 Blenheims and finished with more than 260 Mosquitoes and Mitchells. It flew more than 57,000 operational sorties at a cost of 2671 men killed or missing and 396 wounded.

ORGANIZATION

No. 2 Group
June 1944

- **No. 137 Wing (three squadrons with Boston & Mitchell bombers)**
- **No. 138 Wing (three squadrons with Mosquito bombers)**
- **No. 139 Wing (three squadrons with Mitchell bombers)**
- **No. 140 Wing (three squadrons with Mosquito bombers)**

WINGS OF NO. 2 GROUP (1944)

Wing	Squadrons	Aircraft	Role
No. 137	Nos 88, 226 & 342	Boston & Mitchell	Light bomber
No. 138	Nos 107, 305 & 613	Mosquito	Light bomber
No. 139	Nos 98, 180 & 320	Mitchell	Light bomber
No. 140	Nos 21, 464 & 487	Mosquito	Light bomber

Specifications

Crew: 2

Powerplant: 2 x 1103kW (1480hp) Rolls-Royce Merlin 23 engines

Maximum speed: 595km/h (370mph)

Range: 2744km (1705 miles)

Service ceiling: 10,515m (34,500ft)

Dimensions: span 16.51m (54ft 2in); length 13.08m (42ft 11in); height 5.31m (17ft 5in)

Weight: 9072kg (20,000lb) loaded

Armament: 4 x 20mm (0.8in) cannons in wings, plus 4 x 7.7mm (0.303in) MGs in nose

▲ **de Havilland Mosquito FB.Mk VI**

No. 138 Wing / RAF, 1944

With two Merlin engines and a low-drag airframe built largely of a plywood/balsa/plywood sandwich material for low weight and considerable strength, the Mosquito was very adaptable and fast, and in fighter-bomber form carried four cannon, four MGs, up to 907kg (2000lb) of bombs (half this figure internally) and, in some cases, eight unguided rocket projectiles.

No. 83 Group

1944–45

No. 83 Group was notably important to the British and Commonwealth forces, for it had four reconnaissance and five air observation post squadrons as well as 25 squadrons of fighters and fighter-bombers in nine wings. Five of these wings, with 12 squadrons, were Canadian.

No. 83 Group was formed in the UK on 1 April 1943, and came under the command of Air Vice Marshal W. F. Dickson on 21 March 1943. The group had been planned and created as a primary element of the 2nd TAF. In this formation, No. 83 Group served alongside No. 84 Group, both being supported in the field by the efforts of Air Vice Marshal J. B. Cole-Hamilton's – and from 10 October 1944, Air Vice Marshal C. R. Steele's – No. 85 (Base) Group of 12 squadrons in six wings, as well as essential support elements, including three airfield construction squadrons.

ORGANIZATION

No. 83 Group (Canadian fighter units)
June 1944

- No. 126 Wing (three squadrons with Spitfire fighters)
- No. 127 Wing (three squadrons with Spitfire fighters)
- No. 129 Wing (one squadron with Typhoon fighter-bombers)
- No. 143 Wing (three squadrons with Typhoon fighter-bombers)
- No. 144 Wing (three squadrons with Spitfire fighters)

Ground support role

Nos 83 and 84 Groups were both entrusted with the task of providing provide direct support to British and Canadian forces in the field during the Allied liberation of Europe from 6 June 1944. The two groups played an important part in establishing air superiority and attacking German installations and communications before the commitment of the Allied invasion of Europe in *Overlord*. Air Vice Marshal H. Broadhurst assumed command on 24 March 1944, and by the eve of the landings

No. 83 Group had a strength of 29 squadrons, with fighter and ground-attack units predominating, but including four tactical reconnaissance and five artillery observation post squadrons.

Invasion commitment

The group was committed right from the start of the invasion, and squadrons were moved from the southern part of England to the Normandy lodgement as soon as there were airstrips available for their use. Throughout the northwest Europe campaign, No. 83 Group provided a very high quantity and quality of air support for the formations of the 21st Army Group, including the failed *Market Garden* operations to take and hold bridges over the Rhine river as far north as Arnhem, and the

Specifications

Crew: 1
Powerplant: 1180kW (1580hp) 12-cylinder Merlin Packard 266 engine
Maximum speed: 642km/h (410mph)
Range: 698km (435 miles)
Service ceiling: 12,650m (41,500ft)
Dimensions: span 11.23m (36ft 10in); length 9.47m (31ft 1in); height 3.86m (12ft 8in)
Weight: 3343kg (7370lb) loaded
Armament: 2 x 12.7mm (0.5in) MGs and 2 x 20mm (0.8in) cannons; plus 2 x 114kg (250lb) bombs under each wing

▲ Supermarine Spitfire Mk XVI

No. 74 Squadron / RAF, early 1945

The low-flying Spitfire Mk XVI was the last major production variant of the Merlin-engined Spitfire family, and was powered by a US-made Merlin. Other features were the clear-view canopy and a more pointed vertical tail surface. Some 1054 such aircraft were delivered from October 1944.

▲ de Havilland Mosquito FB.Mk VI

RAF, 1945

The Mosquito FB. Mk VI was the classic fighter-bomber variant of the Mosquito multi-role warplane family, and production totalled 2718 aircraft. The variant entered service in October 1943, and was operated in the day/night intruder and tactical support roles.

Specifications

Crew: 2
Powerplant: 2 x 1103kW (1480hp) Rolls-Royce Merlin 23 engines
Maximum speed: 595km/h (370mph)
Range: 2744km (1705 miles)
Service ceiling: 10,515m (34,500ft)
Dimensions: span 16.51m (54ft 2in); length 13.08m (42ft 11in); height 5.31m (17ft 5in)
Weight: 9072kg (20,000lb) loaded
Armament: 4 x 20mm (0.8in) cannons in wings, plus 4 x 7.7mm (0.303in) MGs in nose

squadrons 'leapfrogged' between airfields close to the frontline in order to keep pace with the Allied advance and exert pressure wherever there was a requirement.

The most important aircraft types flown by No. 83 Group at this time were the Hawker Typhoon Mk I fighter-bomber, Supermarine Spitfire Mk IX fighter and fighter-bomber, and North American Mustang Mk I reconnaissance fighter.

By the time of the German surrender in May 1945, the forward squadrons of No. 83 Group were operating from former *Luftwaffe* airfields in Germany itself, and the group then became part of the British occupation forces before being disbanded, with its squadrons subsequently absorbed into No. 84 Group on 21 April 1946.

WINGS OF NO. 83 GROUP (LATE 1944)

Wing	Squadrons	Aircraft	Role
No. 39	Nos 168, 400 & 414	Mustang	Recce
No. 39	No. 420	Spitfire	Recce
No. 121	Nos 174, 175 & 245	Typhoon	Attack
No. 122	Nos 19, 65 & 122	Mustang	Fighter
No. 124	Nos 181, 182 & 247	Typhoon	Attack
No. 125	Nos 132, 453 & 602	Spitfire	Fighter
No. 126	Nos 401, 411 & 412	Spitfire	Fighter
No. 127	Nos 403, 416 & 421	Spitfire	Fighter
No. 129	No. 184	Typhoon	Attack
No. 143	Nos 438,439 & 440	Typhoon	Attack
No. 144	Nos 441, 442 & 443	Spitfire	Fighter

▲ 'County of Chester' Squadron

No. 610 Squadron of the Royal Auxiliary Air Force flew the Spitfire throughout the war, converting to the superlative Spitfire Mk XIV (pictured here) in January 1944.

No. 84 Group
1944–45

The partner of No. 83 Group within the 2nd TAF, No. 84 Group had AOP and reconnaissance units but was strongest in fighter and fighter-bomber units, with 26 squadrons in eight wings. No. 84 Group also included many units with personnel from occupied countries.

COMMANDED BY Air Vice Marshal L. O. Brown, No. 84 Group partnered No. 83 Group in the 2nd TAF, and had essentially the same role and therefore a comparable composition. The group had three rather than four reconnaissance squadrons with North American Mustang and Supermarine Spitfire aircraft, and two rather than five AOP squadrons with Auster aircraft.The main strength of No. 84 Group lay in its nine wings of Hawker Typhoon ground-attack aircraft (three wings with eight squadrons), Mustang fighter-bombers (one wing with three squadrons) and Spitfire fighters and fighter-bombers (five wings with 15 squadrons). No. 84 Group also included large numbers of 'expatriate' pilots, for its overall order of battle included many foreign squadrons: five Polish, three Czechoslovak, two French, two Norwegian, one Belgian and one New Zealand.

FIGHTER-BOMBER WINGS OF NO. 84 GROUP (JUNE 1944)

Wing	Squadrons	Aircraft	Role
No. 123	Nos 98 & 609	Typhoon	Attack
No. 131	Nos 302, 308 & 317	Spitfire	Fighter
No. 132	Nos 66, 331 & 332	Spitfire	Fighter
No. 133	Nos 129, 306 & 315	Mustang	Fighter
No. 134	Nos 310, 312 & 313	Spitfire	Fighter
No. 135	Nos 222, 349 & 485	Spitfire	Fighter
No. 136	Nos 164 & 183	Typhoon	Attack
No. 145	Nos 329, 340 & 341	Spitfire	Fighter
No. 146	Nos 193, 197, 257 & 266	Typhoon	Attack

ORGANIZATION

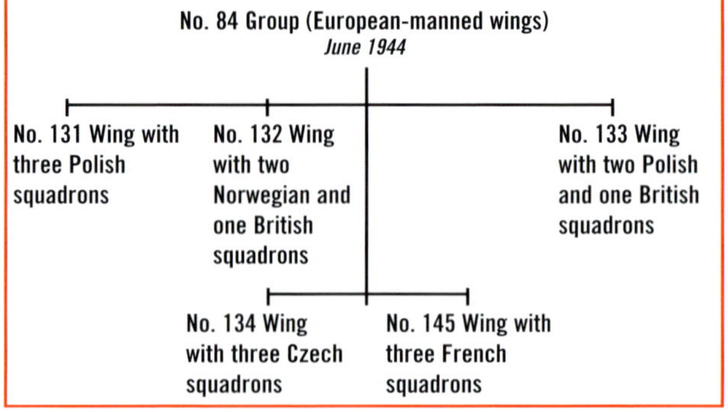

Specifications

Crew: 1
Powerplant: 1081kW (1450hp) V-1650-3 engine
Maximum speed: 690km/h (430mph)
Range: 1215km (755 miles)
Service ceiling: 12,649m (41,500ft)
Dimensions: span 11.27m (37ft); length 9.84m (32ft 4in); height 4.15m (13ft 8in)
Weight: 4173kg (9200lb) loaded
Armament: 4 x 12.7mm (0.5in) MGs

North American Mustang Mk III

No. 133 (Polish) Wing / 2nd TAF / northwest Europe, summer–autumn 1944

This was the personal aeroplane of No. 133 Wing's leader, Wing Commander Stanislaw Skalski, the highest-scoring Polish ace of World War II with at least 19 victories. The wing comprised the British No. 129 Squadron, and the Polish Nos 306 and 315 Squadrons.

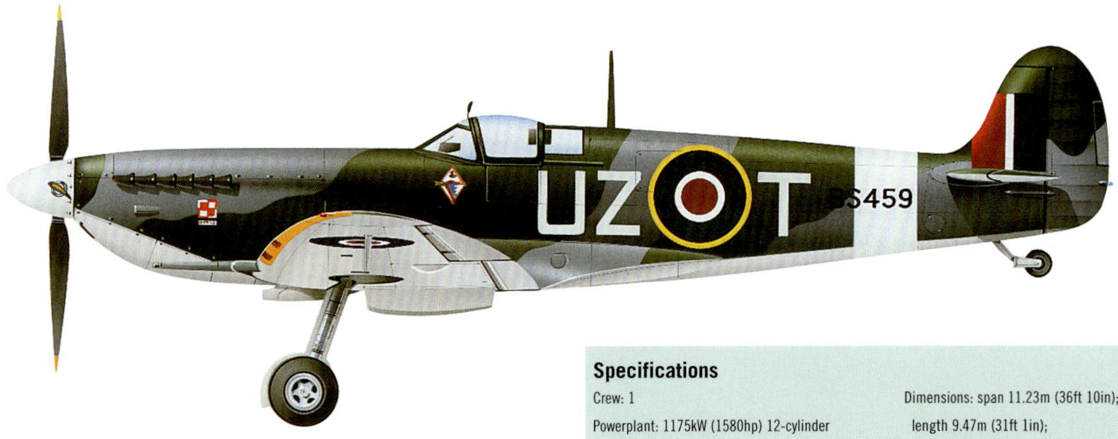

▲ **Supermarine Spitfire Mk IXC**

No. 306 Squadron / No. 133 (Polish) Wing / No. 84 Group, summer 1944

No. 306 Squadron was one of the two Polish-manned squadrons in this wing.

Specifications

Crew: 1	Dimensions: span 11.23m (36ft 10in); length 9.47m (31ft 1in); height 3.86m (12ft 8in)
Powerplant: 1175kW (1580hp) 12-cylinder Rolls-Royce Merlin 66 engine	Weight: 3343kg (7370lb) loaded
Maximum speed: 642km/h (410mph)	Armament: 4 x 7.7mm (0.303in) MGs and 2 x 20mm (0.8in) cannons
Range: 698km (435 miles) on internal fuel tanks	
Service ceiling: 12,650m (41,500ft)	

No. 85 Group

1944–45

No. 85 Group provided a number of support facilities for Nos 83 and 84 Groups. It aided the ground forces directly, but also covered the airfields from which the two more forward groups operated, providing nocturnal coverage against German bombers and fighter-bombers.

COMMANDED BY AIR Vice Marshal J. B. Cole-Hamilton and then, from 10 July, by Air Vice Marshal C. R. Steele, No. 85 Group provided backing for Nos 83 and 84 Groups with airfield construction, beach and balloon squadrons as well as frontline units. It had two wings of day-fighters and four wings of de Havilland Mosquito night-fighters, the latter tasked with protecting the 21st Army Group and all its installations from night attack by the *Luftwaffe*. The night-fighter wings totalled six squadrons, of which three were British, two Canadian and one New Zealand. The squadrons were very effective and, in the period between June 1944 and the end of World War II in Europe in May 1945, were credited with the destruction of some 200 German aircraft.

No. 85 Group's other two combat elements were Nos 141 and 150 Wings, each equipped with day-fighters. No. 141 Wing had three squadrons equipped with Supermarine Spitfire fighters, while

ORGANIZATION

No. 85 Group (night-fighter element)
June 1944

- No.142 wing with two Mosquito squadrons
- No. 147 Wing with one Mosquito squadron
- No. 148 Wing with one Mosquito squadron
- No. 149 Wing with two Mosquito squadrons

ORGANIZATION

No. 85 Group (other elements)
June 1944

- No. 141 Wing with three squadrons
- No. 150 wing with three squadrons
- Airfield Construction Wing with three squadrons
- Beach and Balloon elements with seven squadrons

No. 150 wing also possessed three squadrons. Of these three units, No. 56 Squadron was initially equipped with the Spitfire but converted to the superlative Hawker Tempest fighter on 6 July 1944 and thus flew the same type of warplane as its two sister units, namely Nos 3 and 486 Squadrons.

Initially based on airfields in the southeast of England, No. 85 Wing moved to France as soon as the situation permitted, and then followed in the wake of Nos 83 and 84 Groups as it had longer-range aircraft.

WINGS OF NO. 85 GROUP (1944)

Wing	Squadrons	Aircraft	Role
No. 141	Nos 91, 124 & 322	Spitfire	Fighter
No. 142	Nos 264 & 602	Mosquito	Night-fighter
No. 147	No. 29	Mosquito	Night-fighter
No. 148	No. 409	Mosquito	Night-fighter
No. 149	Nos 410 & 488	Mosquito	Night-fighter
No. 150	No. 56	Spitfire	Fighter
No. 150	Nos 3 & 486	Tempest	Fighter

Push through Northwest Europe

September 1944 – May 1945

By September 1944, the British and Canadian forces of the 21st Army Group had entered Belgium. After the setback of the Arnhem airborne operation and delay as the port of Antwerp was opened, they drove forward into north Germany to reach Hamburg by the end of the war.

The shape of air fighting's future was foreshadowed in October 1944, even as the Allied armies were driving toward Germany's western frontier, by the arrival in service of the Messerschmitt Me 262 turbojet-powered fighter. Available only in small numbers and powered by very unreliable engines, the Me 262 was conceived specifically for this type of powerplant, unlike the slightly earlier straight-winged Arado Ar 234 bomber and the Gloster Meteor that was little more than a conventional fighter recast with two jet engines. The nature of the tactical air war remained essentially

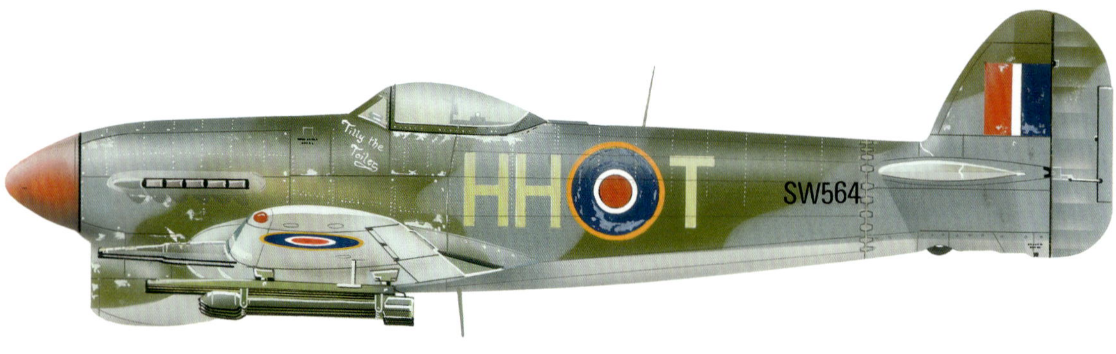

Hawker Typhoon Mk IB

No. 175 Squadron / 2nd TAF, late 1944

No. 175 Squadron received Typhoons in mid 1943, when they became a part of the newly-formed 2nd TAF. The squadron supported the Normandy landings in June 1944, helping to wreck enemy communications networks. The squadron supported Twenty-First Army Group in the ground-attack role through the Low Countries and Germany until the end of the war.

Specifications

Crew: 1

Powerplant: 1685kW (2260hp) Napier Sabre II liquid cooled H-24 in-line piston engine

Maximum speed: 650km/h (405mph)

Range: 980km (610 miles)

Service ceiling: 10,400m (34,000ft)

Dimensions: span 12.67m (41ft 7in); length 9.73m (31ft 11in); height 4.66m (15ft 4in)

Weight: 5170kg (11,400lb) loaded

Armament: 4 x 20mm (0.8in) Hispano-Suiza HS.404 cannons, plus 8 x 27kg (60lb) rockets for ground attack purposes unaltered, however, as the Germans lacked the numbers of aircraft and trained aircrew to make a dent in the Allied support for their ground forces, and they were also further constrained by increasingly scarce fuel reserves. The Allies could therefore operate essentially at will, watchful for only small numbers of German aircraft, with only the Me 262 and a dwindling number of aces flying late-model Focke-Wulf Fw 190s and Messerschmitt Bf 109s posing any real threat.

Two classic fighter-bomber raids of this period were undertaken by de Havilland Mosquito aircraft. On 31 October 1944, 25 Mosquitoes of Nos 21, 464 and 487 Squadrons made a very low-level attack on the Gestapo HQ at Aarhus in Denmark, destroying the building, killing 200 Germans and burning all the files on the Danish resistance. Then, on 31 December, Mosquitoes of No. 627 Squadron made a less successful attack of the same type on the Gestapo HQ in the Norwegian capital, Oslo.

▲ Supermarine Spitfire Mk XVIE

No. 74 Squadron / 2nd TAF, Belgium, late 1944

Optimized for the lower-level role with bombs or rocket projectiles carried externally, the Spitfire Mk XVIE was a definitive member of the Spitfire family, and its 'E' type wing had two 20mm (0.79in) cannon and two 12.7mm (0.5in) MGs.

Specifications

Crew: 1
Powerplant: 1529kW (2050hp) 12-cylinder Griffon 65 engine
Maximum speed: 707km/h (439mph)
Range: 636km (395 miles)
Service ceiling: 12,650m (41,500ft)
Dimensions: span 11.23m (36ft 10in); length 9.47m (31ft 1in); height 3.86m (12ft 8in)
Weight: 3844kg (8475lb) loaded
Armament: 2 x 12.7mm (0.5in) MGs and 2 x 20mm (0.8in) cannons; plus 2 x 114kg (250lb) bombs under each wing

▲ Supermarine Spitfire F.Mk XIVE

No. II Squadron, 1944

The Spitfire XIVE had shortened wings to improve the aircraft's rate of roll at low altitudes. This improved performance in the gound attack role, but gave the wings a 'clipped' appearance.

Specifications

Crew: 1
Powerplant: 1529kW (2050hp) 12-cylinder Griffon 65 engine
Maximum speed: 707km/h (439mph)
Range: 636km (395 miles)
Service ceiling: 12,650m (41,500ft)
Dimensions: span 11.23m (36ft 10in); length 9.47m (31ft 1in); height 3.86m (12ft 8in)
Weight: 3844kg (8475lb) loaded
Armament: 2 x 12.7mm (0.5in) MGs and 2 x 20mm (0.8in) cannons

The *Luftwaffe*'s final raid

The *Luftwaffe* made its final major offensive one day later, sending some 800 aircraft against Allied air bases. The attack gained complete surprise and caused the loss of some 300 Allied aircraft, which could easily be replaced, while the Germans lost a comparable number, including some 200 by their own anti-aircraft guns. The Germans could not replace the few experienced men who were lost, and the tide of the air war flowed even more strongly in favour of the Allies, who were thus untroubled in the air for the rest of the war.

1944 Fighter Wing

By 1944, the Allies had fully absorbed the implications of the Germans' more loosely organized fighter practice, based on the key team of the leader and his wingman. The fighter wing of 1944, therefore, had three or four 16-aircraft squadrons, each operating in four flights, each comprising four aircraft in leader and wingman teams. Combined with superior equipment, such as the Hawker Tempest and late-model Supermarine Spitfire, this arrangement gave the British the tactical advantage.

Squadron 1, Flight 1

Flight 2

Flight 3

Flight 4

Squadron 2, Flight 1

Flight 2

Flight 3

Flight 4

Squadron 3, Flight 1

Flight 2

Flight 3

Flight 4

Squadron 4, Flight 1

Flight 2

Flight 3

Flight 4

RAF Coastal Command

1943–45

Established in July 1936 to succeed the RAF's Coastal Area, Coastal Command protected the UK from naval attack, sought to destroy German surface raiders and U-boats while covering convoys and took the air war to Germany's coastal shipping. By 1943, it was winning its war.

COASTAL COMMAND was controlled operationally by the Admiralty for much of the war to maximize the UK's sea and air response to German surface and submarine threats. Within this context it also collaborated with the shore-based elements of the Fleet Air Arm in the UK, and with the Royal Navy's carrier forces in home waters and the North Atlantic.

ORGANIZATION

Coastal Command, UK (Air Marshal Sir John Slessor)
June 1944

No. 15 Group	No. 16 Group	No. 18 Group	No. 19 Group

Coastal Command structure

In September 1939, the command had four groups, three of them operational round the coast of the UK: No. 16 Group in the eastern half of the English Channel and the southern half of the North Sea, No. 18 Group in the rest of the North Sea and areas to the north and west of Scotland, and No. 15 Group in the rest. In February 1941, No. 19 Group became active between Nos 18 and 15 Groups. In November 1940, No. 200 Group at Gibraltar was transferred to the control of Coastal Command from RAF Mediterranean, and in December 1941 No. 200 Group became RAF Gibraltar. Coastal Command also had in Iceland No. 30 Wing, which in July 1941 became RAF Iceland, and from mid-1943 No. 247 Group in the Azores.

Command aircraft

The command's equipment was initially inferior, especially where land-based aircraft were concerned.

Specifications

Crew: 3
Powerplant: 2 x 742kW (995hp) Bristol Mercury XV nine cylinder engine
Maximum speed: 266km/h (428mph)
Range: 2350km (1460 miles)
Service ceiling: 6705m (22,000ft)
Dimensions: span 17.17m (56ft 4in); length 12.98m (42ft 7in); height 3.90m (12ft 9in)
Weight: 6804kg (15,000lb) loaded
Armament: 1 x 7.7mm (.303in) MG in lead edge of port wing, 2 x 7.7mm (.303in) trainable MGs in dorsal turret, 2 x 7.7mm (.303in) forward-firing MGs in undernose blister position

▲ **Bristol Blenheim Mk IVF**

No. 248 Squadron / Coastal Command

No. 248 Squadron was one of 11 Coastal Command squadrons that flew the Blenheim Mk IVF in the medium-range fighter role, roaming out into the waters off the British Isles to watch for Germans surface raiders and U-boats, and to destroy German bomber and reconnaissance aircraft. They also attacked light craft in coastal waters.

▲ **Rocket attack**

The Beaufighter was a lethal maritime strike aircraft. Here, a Mk. X from No. 455 Squadron releases all eight rocket at once against a seaborne target.

MAJOR COASTAL COMMAND AIRCRAFT TYPES		
Aircraft	**Role**	**No. of sqds**
Avro Anson	Short-range recce	21
Bristol Beaufort	Torpedo	6
Bristol Beaufighter	Attack	13
Consolidated Catalina	Long-range recce	9
Consolidated Liberator	Long-range recce	12
de Havilland Mosquito	Attack	7
Handley Page Hampden	Torpedo & weather	7
Lockheed Hudson	Medium-range recce	17
Short Sunderland	Long-range recce	14
Vickers Warwick	Rescue & recce	10
Vickers Wellington	Medium-range recce	14

However, as the war progressed and the threat to convoys increased, Coastal Command received a higher priority, and by 1943 was well equipped in terms of its flying boats and land-based reconnaissance bombers.

Flying boats included the Short Sunderland and Consolidated Catalina, while the land-based types included the Vickers Wellington, Boeing B-17 Fortress and Consolidated Liberator. The first and the last two offered the ability to roam deep into the Arctic and Atlantic Oceans.

The arrival of the de Havilland Mosquito freed the Bristol Beaufighter for Coastal Command use in the shorter-range role. The Beaufighter became a very useful short-range aeroplane, operating with rockets and depth charges against U-boats in the Bay of Biscay. The type was also used in attacks on German shipping, even attacking German flak vessels. The Beaufighter was later supplemented in this role by the Mosquito fighter-bomber.

By the start of 1943, improvements in tactics and the introduction of electronic aids vastly improved the command's capability, and as shipping losses declined the U-boat kill rate rose. It was not so much the number sunk as the constant harassment they could inflict which made the aircraft effective: if the U-boats approached to make contact in daylight, they risked interception, and even night-time attacks were perilous.

During the war, Coastal Command flew more than 240,000 sorties, sank 212 U-boats and destroyed 478,000 tons of shipping, in the process losing 1777 aircraft and as many as 10,875 men.

Coastal Command strike wings

1944–45

The units that took the anti-ship war to the Germans were Coastal Command's strike wings, mostly located on the east coast of Scotland. They crossed the North Sea in search of coastal shipping in the Leads of occupied Norway's long western shore.

THE TWO AIRCRAFT types generally associated with the strike wings were the Bristol Beaufighter and de Havilland Mosquito twin-engined heavy fighter-bombers, the former armed with a torpedo and rockets, and the latter with bombs, rockets and occasionally a 57mm (2.24in) gun.

On 1 September 1944, No. 18 Group assumed control of RAF Banff, and the No. 144 Squadron's

ORGANIZATION

Banff Strike Wing, Coastal Command
September 1944 – May 1945

Nos 143, 144, 235 & 248 Squadrons, RAF	No. 404 Squadron, RCAF	No. 333 (Norwegian) Squadron	No. 445 (Australian) Squadron

▲ **Banff Strike Wing**

No. 404 Squadron, RCAF, flew the Bristol Beaufighter from September 1942, receiving these Beaufighter TF.Mk X machines in September 1943.

Specifications

Crew: 2
Powerplant: 2 x 1320kW (1770hp) Bristol Hercules XVII radial piston engines
Maximum speed: 488km/h (303mph)
Range: 2366km (1470 miles)
Service ceiling: 4570m (15,000ft)
Dimensions: span 17.63m (57ft 10in); length 12.70m (41ft 8in); height 4.83m (15ft 10in)
Weight: 11,431kg (25,200lb) loaded
Armament: 4 x 20mm (0.8in) cannon and 6 x 7.7mm (.303in) MGs in wings; 1 x 7.7mm (.303in) Vickers MG in dorsal turret; 1 x torpedo

▲ Bristol Beaufighter TF.Mk X

No 489 (New Zealand) Squadron / RAF Coastal Command, UK, 1944–45

Sturdy and fast at low altitude, the Beaufighter TF.Mk X carried an air-launched torpedo under its fuselage, or rocket projectiles under its outer wing panels.

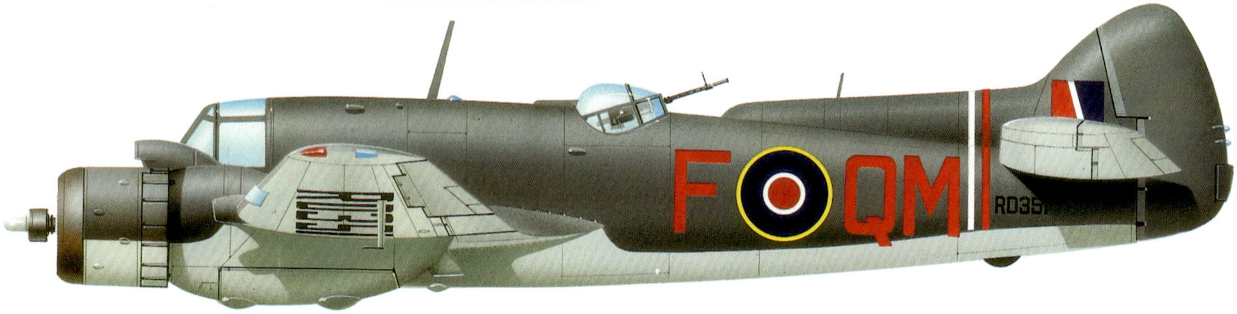

Specifications

Crew: 2
Powerplant: 2 x 1320kW (1770hp) Bristol Hercules XVII radial piston engines
Maximum speed: 488km/h (303mph)
Range: 2366km (1470 miles)
Service ceiling: 4570m (15,000ft)
Dimensions: span 17.63m (57ft 10in); length 12.70m (41ft 8in); height 4.83m (15ft 10in)
Weight: 11,431kg (25,200lb) loaded
Armament: 4 x 20mm (0.8in) cannon and 6 x 7.7mm (.303in) MGs in wings; 1 x 7.7mm (.303in) Vickers MG in dorsal turret

▲ Bristol Beaufighter TF.Mk X

No. 254 Squadron / North Coates Strike Wing / RAF Coastal Command, UK, 1944

Late-service examples of the Beaufighter TF.Mk X were operated in a two-tone grey camouflage scheme optimized for use in the sky and sea conditions and colours prevalent in the North Sea and Norwegian Sea.

Beaufighters and Nos 235 and 248 Squadrons' Mosquitoes arrived, together with 'P' flight of No. 333 (Norwegian) Squadron (with Mosquitoes to act as outriders on the basis of its knowledge of the Norwegian coast), and finally No. 404 (Canadian) Squadron's Beaufighters. The commanding officer was Wing Commander F. W. Pierce.

First raid

The wing's first major effort was made on 14 September, when 25 Mosquitoes of Nos 235 and 248 squadrons, along with four of No. 248 Squadron's Mosquito Mk XVIIIs with the 57mm (2.24in) gun, flew a 'Rover' armed patrol with 19 Beaufighters of Nos 144 and 404 Squadrons. The

ORGANIZATION

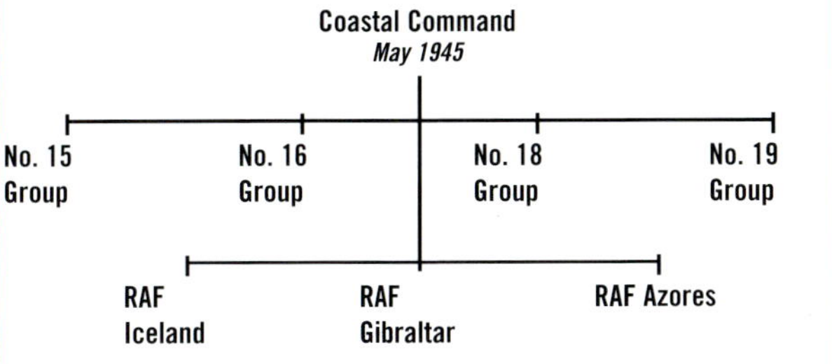

COASTAL COMMAND SQUADRON TOTALS (MAY 1945)

Group	Anti-submarine squadrons/strength	Anti-ship squadrons/strength
No. 15	7/100	0
No. 16	2.5/28	5/94
No. 18	8/105	9.5/180
No. 19	13/183	0
Iceland	2/30	0
Gibraltar	2/35	0
Azores	2/30	0

mission found four motor vessels with two escorts off Kristiansund and attacked, hits being scored on all vessels and fires started. The 268-tonne (264-ton) flagship *Sulldorf* was sunk and the 3376-tonne (3323-ton) merchant vessel *Iris* was damaged. The flak was intense, which forced one Beaufighter to ditch offshore.

On 19 September, another Rover with 21 Beaufighters and 11 Mosquitoes attacked a convoy of three ships near Askevold, sinking two of the merchantmen, one of 1389 tonnes (1367 tons) and the other of 3129 tonnes (3080 tons), for the loss of one No. 144 Squadron Beaufighter. On 21 September, 21 Beaufighters escorted by 17 Mosquitoes attacked and sank two merchant vessels, and five days later 16 Mosquitoes sank the *Biber* and damaged the *Storfsund* off the Hjeltefjord. Late in the month, the Mosquitoes were modified to carry eight rockets. On 21 October, Mosquitoes of Nos 235 and 248 Squadrons and Beaufighters of No. 404 Squadron attacked shipping in Haugesend harbour, sinking the 1954-tonne (1923-ton) *Eckenheim* and 1455-tonne (1432-ton) *Vestpa* for the loss of one aeroplane.

North Sea raids

On 5 December, Mosquitoes of Nos 143, 235 and 248 Squadrons flew a major attack on shipping in the Nordgulenfjord, which resulted in four German vessels being damaged. These were the *Ostland*, *Tucuman*, *Magdalena* and *Helene Russ*. During this action, two Mosquitoes were badly hit, one of No. 143 Squadron reaching Sumburgh in the Shetlands on one engine and crash-landing, and the

Specifications

Crew: 2
Powerplant: 2 x 1230kW (1650hp) Bristol Hercules VI radial piston engines
Maximum speed: 488km/h (303mph)
Range: 2366km (1470 miles)
Service ceiling: 4570m (15,000ft)
Dimensions: span 17.63m (57ft 10in); length 12.70m (41ft 8in); height 4.83m (15ft 10in)
Weight: 11,431kg (25,200lb) loaded
Armament: 4 x 20mm (0.8in) cannon and 6 x 7.7mm (.303in) MGs in wings; 1 x 7.7mm (.303in) Vickers MG in dorsal turret; 8 x 41kg (90lb) rockets

▲ Beaufighter Mk VI

RAF Coastal Command, 1944

Features of all but the first Beaufighters were the dihedralled rather than flat tailplane and, as development continued, enlargement of the vertical tail area with a long dorsal fillet.

ARMAMENT OF SELECTED COASTAL COMMAND AIRCRAFT

Type	Guns	Bombs	Torpedo/rocket
Avro Anson I	2 x 7.7mm (0.3in)	163kg (360lb)	0
Bristol Beaufighter VI	4 x 20mm (0.8in) & 4x 7.7mm (0.3in)	2 x 113kg (250lb)	1/8
de Havilland Mosquito VI	4 x 20mm (0.8in) & 4x 7.7mm (0.3in)	4 x 227kg (500lb)	0/8
Consolidated Catalina I	4 x 7.7mm (0.3in)	907kg (2000lb)	0/0
Short Sunderland V	2 x 12.7mm (0.5in) & 12 x 7.7mm (0.3in)	907kg (2000lb)	0/0
Vickers Wellington III	8 x 7.7mm (0.3in)	2040kg (4500lb)	0/0

other of No. 248 Squadron being shot down. The latter aeroplane was seen to make an attack on a large, heavily armed ocean-going tug, and was on fire before making the attack. It later crashed into the sea. During the attack in the Nordgulenfjord, all aircraft encountered intense light flak from the whole of the eastern end of the anchorage, especially from the northern shore, some of the guns being positioned several hundred feet up the mountain side. Five other aircraft also had to land away from base, three of them on one engine; another four suffered damage.

On 7 December, a mixed sortie by 25 Mosquitoes from Banff and 40 Beaufighters from Dallachy, escorted by 12 Mustangs of No. 315 (Polish) Squadron, attacked the fighter airfield at Gossen in Norway. The formation was attacked by 12 Messerschmitt Bf 109 and Focke-Wulf Fw 190 fighters, and in the air battle that followed No. 315 Squadron claimed four Bf 109s downed. One Mustang, one Beaufighter and two Mosquitoes were lost.

On 10 December, an attack by Nos 143, 235 and 248 Squadrons' Mosquitoes attacked shipping in the Flekkefjord, and sank the 1508-tonne (1485-ton) *Gijdrun*. No aircraft were lost during the attack.

On 16 December, an attack force of Nos 143, 235 and 248 Squadrons attacked shipping at Malloy and Kraakbellesund. At the former the 5775-tonne (5684-ton) *Ferndale* was sunk, while at the latter the tug *Parat* was also sunk. During these actions, two Mosquitoes were lost. One was an aeroplane of No. 248 Squadron, which was hit by flak. This machine managed a controlled ditching. Both members of the crew were seen to leave the aeroplane and climb into their dinghy. A Vickers Warwick air-sea rescue aeroplane dropped a Lindholme dinghy. Up to the time when escorting aircraft had to return to base as they ran short of fuel, the two men were seen to be sitting up in the dinghy, but despite an intensive search of the area they were never found.

Thus the effort continued right into May 1945.

Coastal Command Squadron (1943–45)

By the middle period of World War II, the organization of Coastal Command had settled into the workable pattern suggested by experience in the more fraught early stages of the war. The attack squadrons of Coastal Command, flying aircraft such as the Bristol Beaufighter and de Havilland Mosquito (below) in the second half of the war, thus had 12 aircraft subdivided into four three-aircraft flights.

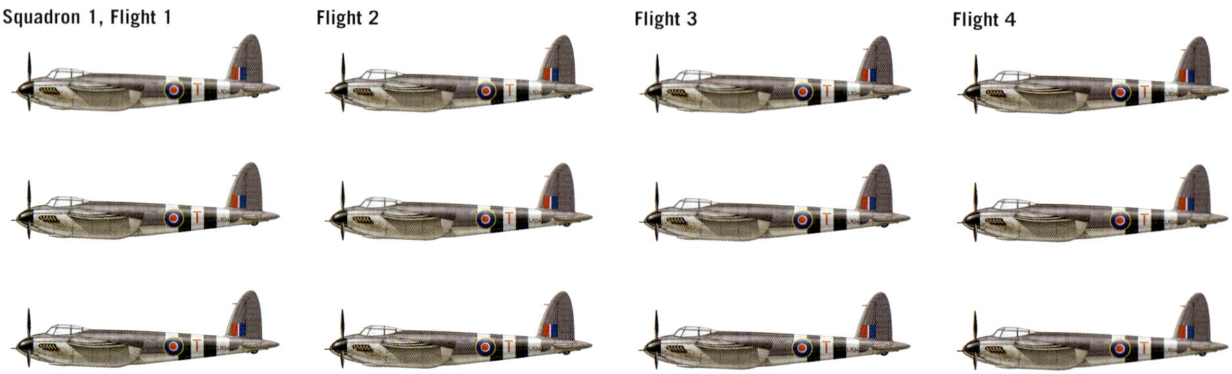

Maritime fighters
1943–45

The fighters serving with Coastal Command and the Fleet Air Arm fell into two basic forms as large twin-engined strike fighters used only by Coastal Command, and smaller single-engined fighters used only by the FAA for both the carrierborne and land-based roles.

THE PRIMARY STRIKE fighters of RAF Coastal Command were a pair of substantial twin-engined aircraft, namely the Bristol Beaufighter and de Havilland Mosquito. The Beaufighter had been designed as a heavy fighter possessing a high degree of conceptual similarity with the Blenheim light bomber and Beaufort torpedo-bomber, and was powered by two Bristol Hercules radial engines.

ORGANIZATION

Coastal Command Strike Wings
November 1942–May 1945

North Coates Strike Wing (Nos 143, 236 & 254 Squadrons)	Langham and Dallachy Strike Wing (Nos 455, 489 Squadrons)	Strubby and Dallachy Strike Wing (Nos 144 & 404 Squadrons)	Banff Strike Wing (Nos 143, 235, 248 & 333 Squadrons)

Anti-shipping role

Entering service in September 1940, the Beaufighter was developed in four main variants for Coastal Command: 1) the initial Beaufighter Mk IC; 2) the Beaufighter Mk VIC with provision for an underfuselage torpedo; 3) the Beaufighter TF.Mk X development of the Mk VIC with greater power, radar and provision for underwing rockets; 4) the Beaufighter TF.Mk XI with no torpedo capability. Beaufighter-equipped strike wings became the scourge of German shipping in the North Sea and English Channel; in two days during April 1945, Beaufighters sank five U-boats.

New fighter

Entering service in May 1943 in its fighter form, the de Havilland Mosquito was powered by two Rolls-Royce Merlin V-12 engines, and was used by Coastal Command in two primary variants. These entered

Specifications

Crew: 2
Powerplant: 2 x 1320kW (1770hp) Bristol Hercules XVII radial piston engines
Maximum speed: 488km/h (303mph)
Range: 2366km (1470 miles)
Service ceiling: 4570m (15,000ft)
Dimensions: span 17.63m (57ft 10in); length 12.70m (41ft 8in); height 4.83m (15ft 10in)
Weight: 11,431kg (25,200lb) loaded
Armament: 4 x 20mm (0.8in) cannon and 6 x 7.7mm (.303in) MGs in wings; 1 x 7.7mm (.303in) Vickers MG in dorsal turret; 2 x 113kg (250lb) bombs

▲ **Bristol Beaufighter TF.Mk X**

No. 455 (Australian) Squadron / Langham & Dallachy Strike Wing, mid-1944

One of the favourite weapons of Beaufighter strike squadrons was the 76mm (3in) rocket fitted with either an 11kg (25lb) solid armour-piercing or 27kg (60lb) explosive-filled semi-armour-piercing warhead. The projectile's accuracy was limited, but the full salvo of 16 such weapons was truly devastating.

service in May 1944, and were the Mosquito FB.Mk VI fighter-bomber (essentially similar to that used by Fighter Command with machine-gun and cannon armament as well as bombs or rocket projectiles), and the Mosquito FB.Mk XVIII, known as the 'Tsetse' for its sting in the offensive role, as the standard four 20mm (0.79in) cannon were replaced by a single 57mm (2.24in) adapted anti-tank gun.

Fleet Air Arm – North Atlantic
1943–45

Using increasing numbers of aircraft delivered by the United States as the primary equipment for its growing force of escort carriers, the Fleet Air Arm became a vital weapon in the battle of the North Atlantic against German U-boats.

During 1943, the Royal Navy's escort carrier strength increased rapidly, and this allowed most major convoys to be escorted by at least one such vessel. Built to mercantile rather than naval standards, and smaller than the fleet's large and light carriers, the escort carrier could be produced fairly quickly, and it was less of a loss, although still a blow, when such a vessel was sunk.

Several types of warplane saw service on the escort carriers, but the three most important were all products of the Grumman company. These were the

MAJOR DELIVERIES OF GRUMMAN AIRCRAFT TO THE ROYAL NAVY

Aircraft	Role	Number	US Navy
Wildcat I, II & III	Fighter	100, 90 &10	F4F-3A
Wildcat IV	Fighter	200	F4F-4
Wildcat V & VI	Fighter	312 & 370	FM-1 & -2
Hellcat I	Fighter-bomber	252	F6F-3
Hellcat II	Fighter-bomber	930	F6F-5
Avenger I	Torpedo bomber	402	TBF-1
Avenger II & III	Torpedo bomber	334 & 222	TBM-1C & -3

Vought Corsair Mk I

No. 1835 Squadron / Fleet Air Arm / Brunswick, Nova Scotia, late 1943

The Royal Navy was an early operator of the Corsair, a highly effective land-based and carrierborne fighter well able to tackle surface as well as air targets.

Specifications

Crew: 1
Powerplant: 1678kW (2250hp) Pratt & Whitney R-2800-18W 18-cylinder radial engine
Maximum speed: 718km/h (446mph)
Range: 2511km (1560 miles)
Service ceiling: 12,650m (41,5000ft)
Dimensions: span 12.49m (41ft); length 10.27m (33ft 8in); height 4.50m (14ft 9in)
Weight: 6149kg (13,555lb) loaded
Armament: 6 x 12.7mm (0.5in) forward-firing MGs in wings

Wildcat (originally Martlet) fighter, the Hellcat fighter-bomber, and the Avenger torpedo and level bomber. All were powered by very reliable air-cooled radial engines, and while the Wildcat was useful primarily for patrol round the convoy and attacks on U-boats with its machine-gun armament, the Hellcat offered greater performance as well as the ability to carry two 454kg (1000lb) bombs or eight rocket projectiles, both effective anti-submarine weapons. The Avenger offered a different level of capability with much longer endurance and the ability to carry one torpedo, or 907kg (1000lb) of bombs or rockets.

Specifications

Crew: 1
Powerplant: 1491kW (2000hp) Pratt & Whitley R-2800-10W radial piston engine
Maximum speed: 597km/h (371mph)
Range: 1674km (1040 miles)
Service ceiling: 11,186m (36,700ft)
Dimensions: span 13.06m (42ft 10in); length 10.24m (33ft 7in); height 4.39m (33ft 7in)
Weight: 4178kg (9212lb) loaded
Armament: 6 x 12.7mm (0.5in) wing-mounted MGs plus 6 x 27kg (60lb) rockets or 2 x 454kg (1000lb) bombs

▲ Grumman Hellcat Mk I

No. 800 Squadron / Fleet Air Arm / embarked on HMS* Emperor, *1944

The potent carrierborne fighter known to the British as the Hellcat Mk I was the Lend-Lease counterpart of the US Navy's F6F-3 Hellcat, the first production variant of this successor to the Wildcat. Deliveries to the UK amounted to 252 aircraft.

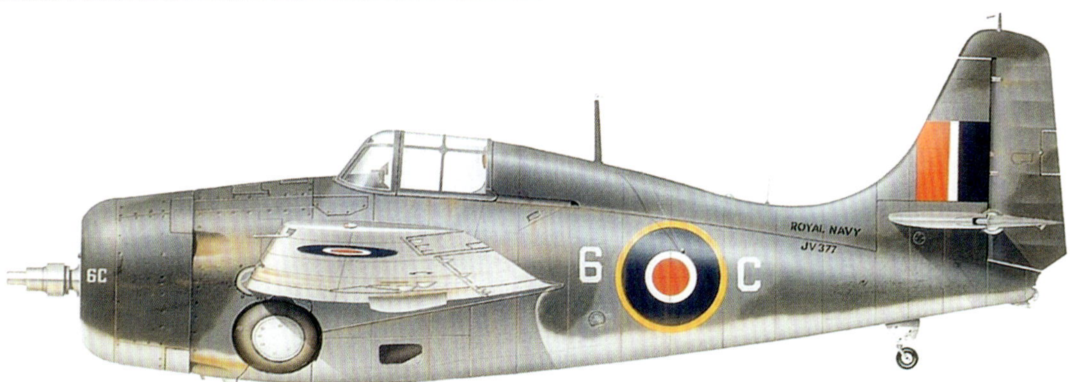

▲ Grumman Wildcat Mk VI

No. 882 Squadron / Fleet Air Arm

Wildcat Mk VI was the British designation for the FM-2, the last variant of the F4F Wildcat family and built by General Motors rather than Grumman. Some 370 FM-2s were transferred to the UK under the terms of the Lend-Lease Act.

Specifications

Crew: 1
Powerplant: 1007kW (1350hp) Pratt & Whitney R-1830-86 14-cylinder engine
Maximum speed: 507km/h (315mph)
Range: 1851km (1150 miles)
Service ceiling: 10,900m (35,700ft)
Dimensions: span 11.58m 38ft); length 8.76m (28ft 8in); height 2.81m (9ft 3in)
Weight: 3607kg (7952lb) loaded
Armament: 6 x 12.5mm (0.5in) MGs in wings; 2 x 113kg (249lb) bombs

LOU IV
413410
C
S
E2
A

Chapter 3

United States

In December 1941, when the USA was drawn into World War II by the Japanese attack on Pearl Harbor, the US Army Air Forces (USAAF) had 4002 combat aircraft. By August 1945, at the time of Japan's surrender, this strength had increased to 39,192, while over the same period non-combat strength had grown from 12,297 to 63,475. Within the combat totals, the USAAF's fighter strength rose from 2170 to 16,799 aircraft. This huge growth in less than five years, was made possible only by the USA's timely prewar decision to start a programme of massive rearmament, and also to boost industrial production to meet not only its own requirements but those of the European nations fighting the Germans.

Cadillacs of the sky

P-51D Mustang fighters of the 357th Fighter Squadron, 361st Fighter Group, fly above England on their way to Germany, 1944.

Early organization
1942–43

The strategic air formation that the United States based in the UK was the Eighth Air Force. This formation grew from 1942 totals of four fighter, one medium bomber and 10 heavy bomber groups to March 1945 figures of nine fighter and 43 heavy bomber groups.

The American concept of how a major enemy should be defeated was already well established before the USA was drawn into World War II by the Japanese attack of 7 December 1941 on Pearl Harbor in the Hawaiian Islands. This event enlarged World War II by a huge extent, and what was now truly a global conflict was further widened by Germany's declaration of war on the USA.

US military doctrine had long emphasized the paramount importance of a direct attack on a major enemy in massive strength and by the most direct route. This was primarily the task of the US Army, which would enjoy the benefits of powerful tactical air support from the USAAF, but a further role was planned for air power. This was the destruction, or at least the severe degradation, of the enemy's ability to wage or sustain a modern war by the systematic application of strategic air power.

The target of this strategic air power was the destruction of the German war industries by precision daylight bombing. The main targets for this

ORGANIZATION

Eighth Air Force
November 1942

VIII Bomber Command	VIII Fighter Command	VIII Ground Air Support Command	VIII Air Service Command

Mustang ace

Fighter leader Captain Don Gentile scored 15.5 kills in the North American P-51 Mustang, flying with the 336th Fighter Squadron of the Eighth Air Force's 4th Fighter Group.

Supermarine Spitfire Mk VB

309th Fighter Squadron / 31st Fighter Group, late 1942

To boost the number of capable combat aircraft available to it, the USAAF used small numbers of British aircraft, including this Spitfire Mk VB, which came with ex-'Eagle' squadron personnel, and later passed to the 67th Observation Group.

Specifications

Crew: 1
Powerplant: 1096kW (1470hp) Rolls-Royce Merlin 50 engine
Maximum speed: 594km/h (369mph)
Range: 1827km (1135 miles)
Service ceiling: 11,125m (36,500ft)
Dimensions: span 11.23m (36ft 10in); length 9.12m (29ft 11in); height 3.02m (9ft 11in)
Weight: 2911kg (6417lb) loaded
Armament: 4 x 7.7mm (0.303in) MGs and 2 x 20mm (0.8in) cannons in wings

effort were most obviously key industrial targets, such as the factories in which aircraft and tanks were made, and – most importantly of all – any chokepoints in the industrial process, such as the few ball-bearing manufacturing facilities, whose destruction would reduce Germany's capacity for the manufacture of engines for aircraft, tanks and submarines, and even the machine tools on which all modern industrial processes were dependent. Another major target area was the communications needed to move raw material to the factories and deliver completed weapons to the armed forces.

PRODUCTION TOTALS FOR US FIGHTERS OPERATED OVER EUROPE

Aircraft	Type	Entered service	Number built
Bell P-39 Airacobra	Fighter-bomber	1941	9590
Bell P-63 Kingcobra	Fighter-bomber	1943	3505
Lockheed P-38 Lightning	Fighter	1941	9393
North American P-51 Mustang	Fighter	1942	15,469
Republic P-47 Thunderbolt	Fighter	1943	15,634

Supermarine Spitfire Mk VB

334th Fighter Squadron / 4th Fighter Group, early 1943

The 4th Fighter Group was activated on 12 September 1942, as the RAF's Nos 71, 121 and 133 'Eagle' Squadrons (personnel and aircraft) were transferred to become the 334th, 335th and 336th Fighter Squadrons. The 4th Fighter Group claimed the highest total of German aircraft destroyed in the air and on the ground, and on 28 July 1943 was the first Eighth Air Force fighter group to fly over Germany itself.

Specifications

Crew: 1
Powerplant: 1096kW (1470hp) Rolls-Royce Merlin 50 engine
Maximum speed: 594km/h (369mph)
Range: 1827km (1135 miles)
Service ceiling: 11,125m (36,500ft)
Dimensions: span 11.23m (36ft 10in); length 9.12m (29ft 11in); height 3.02m (9ft 11in)
Weight: 2911kg (6417lb) loaded
Armament: 4 x 7.7mm (0.303in) MGs and 2 x 20mm (0.8in) cannons in wings

Republic P-47C Thunderbolt

334th Fighter Squadron / 4th Fighter Group, Debden, March 1943

After flying the Spitfire Mk V fighters it brought from service in the RAF's No. 71 'Eagle' Squadron, the 334th Fighter Squadron converted to P-47C heavy fighters.

Specifications

Crew: 1
Powerplant: 1891kW (2535hp) Pratt & Whitney R-2800-59W Double Wasp
Maximum speed: 697km/h (433mph)
Range: 3060km (1900 miles) with drop tanks
Service ceiling: 12,495m (41,000ft)
Dimensions: span 12.42m (40ft 9in); length 11.02m (36ft 2in); height 4.47m (14ft 8in)
Weight: 7938kg (17,500lb) maximum take-off
Armament: 8 x 12.7mm (0.5in) MGs in wings

Strategic bomber escort
1943–45

The Americans initially believed that the Boeing B-17 heavy bomber's high performance and defensive armament would permit daylight precision bombing without the need for fighter escort. Events proved them wrong, and the search was launched for a long-range escort.

THE USA BELIEVED that the way to win the war in Europe was by direct attack on Germany's primary strength by the ground forces, whose task would be aided by the air-delivered destruction of Germany's industrial and military production capabilities. This was the task of the strategic bomber forces which, the USAAF thought, had the right weapon in the form of the Boeing B-17 four-engined heavy bomber with the advanced Norden bomb sight. This was intended to allow bombing by day from high altitude with the accuracy to destroy key German industrial facilities.

ESCORT FIGHTERS OF THE EIGHTH AIR FORCE

Type	Armament	Standard range	Range with drop tanks
Lockheed P-38L	1 x 20mm (0.79in) 4 x 12.7mm (0.5in)	1890km (1175 miles)	4185km (2600 miles)
North American P-51D	6 x 20mm (0.79in)	1530km (950 miles)	3700km (2300 miles)
Republic P-47D	8 x 12.7mm (0.5in)	1270km (790 miles)	2775km (1725 miles)

'Box' formation

The bombers would fly at more than 7620m (25,000ft) in three-dimensional 'box' formations, these designed so that each bomber was covered by the guns of surrounding bombers. The 'boxes' were supposed to be relatively immune from interception except by small numbers of fighters, which would be defeated by the bombers' massed 12.7mm (0.5in) trainable machine guns.

It was a vain hope, as the Eighth Air Force's first forays into German air space soon revealed. Radar allowed the Germans to scramble fighters early enough to reach high altitude in some numbers. The German fighter pilots quickly established that the

▲ **'Ferocious Frankie'**

This bomb-loaded P-51D Mustang was the regular mount of Major Wallace Hopkins, commanding officer of the 374th Fighter Squadron, 361st Fighter Group.

bombers were most vulnerable to head-on attacks, and that it was possible to break into these 'boxes'. What the bombers needed was escort by fighters with the agility and firepower, and with the range to accompany the bombers to and from their targets.

The Republic P-47 and Lockheed P-38 provided a partial solution, but lacked the agility to cope with the superior German fighters (such as the Fw 190) and the range to remain with the bombers right through the mission. The solution appeared in the form of the P-51 Mustang, which was in itself a superb fighter, had considerable internal fuel capacity, and was finally fitted with drop tanks for prodigious range. The tanks were dropped before combat took place, the fighters then returning home after the sortie on internal fuel.

▲ **North American P-51B Mustang**

354th Fighter Squadron / 355th Fighter Group, spring 1944

The 355th Fighter Group reached Staple Morden, its home for the rest of the war, in July 1943, and its 354th, 357th and 358th Fighter Squadrons were equipped with the P-47D until February 1944. The group claimed more strafing victories than any other group of the Eighth Air Force.

Specifications

Crew: 1
Powerplant: 1044kW (1400hp) Rolls-Royce Merlin V-1650-3 12-cylinder engine
Maximum speed: 690km/h (430mph)
Range: 3540km (2200 miles)
Service ceiling: 12,649m (41,500ft)
Dimensions: span 11.27m (37ft); length 9.84m (32ft 4in); height 4.15m (13ft 8in)
Weight: 4173kg (9200lb) loaded
Armament: 6 x 12.7mm (0.5in) MGs

▲ **Lockheed P-38L-5-LO Lightning**

55th Fighter Squadron / 20th Fighter Group / King's Cliffe, mid-1944

The 57th, 77th and 79th Fighter Squadrons of the 20th Fighter Group took up station at King's Cliffe in August 1943 with the P-38H and P-38J variants of the Lightning, but converted to the P-51C in December 1944.

Specifications

Crew: 2
Powerplant: 2 x 1194kW (1600hp) Allison V-1710-111/113
Maximum speed: 666km/h (414mph)
Range: 4184km (2600 miles)
Service ceiling: 13,410m (20,000ft)
Dimensions: span 15.85m (52ft); length 11.53m (37ft 10in); height 2.99m (9ft 10in)
Weight: 9798kg (21,600lb) maximum takeoff
Armament: 1 x 20mm (0.8in) cannon and 4 x 12.7mm (0.5in) MGs in nose, plus 2 x 907kg (2000lb) bombs

Specifications

Crew: 1
Powerplant: 1081kW (1450hp) Rolls-Royce Merlin V-1650-3 12-cylinder engine
Maximum speed: 690km/h (430mph)
Range: 1215km (755 miles)
Service ceiling: 12,649m (41,500ft)
Dimensions: span 11.27m (37ft); length 9.84m (32ft 4in); height 4.15m (13ft 8in)
Weight: 4173kg (9200lb) loaded
Armament: 4 x 12.7mm (0.5in) MGs

▲ North American P-51B Mustang

374th Fighter Squadron / 361st Fighter Group, Bottisham, mid-1944

The 361st Fighter Group arrived at Bottisham with its 374th, 375th and 376th Fighter Squadrons in November 1943. The group first flew the P-47D, but converted to the P-51 in May 1944. Later homes were Little Walden, St Dizier, Chievres and, finally, Little Walden once more.

Specifications

Crew: 1
Powerplant: 1264kW (1695hp) Packard Merlin V-1650-7 V-12 piston engine
Maximum speed: 703km/h (437mph)
Range: 3347km (2080 miles)
Service ceiling: 12,770m (41,900ft)
Dimensions: span 11.28m (37ft); length 9.83m (32ft 3in); height 4.17m (13ft 8in)
Weight: 5488kg (12,100lb) loaded
Armament: 6 x 12.7mm (0.5in) MGs in wings, plus up to 2 x 454kg (1000lb) bombs

▲ North American P-51D Mustang

362nd Fighter Squadron / 357th Fighter Group, Leiston, mid-1944

Always equipped with Mustang variants, the 357th Fighter Group began its life in the Eighth Air Force at Raydon during November 1943, but two months later moved to Leiston with its 362nd, 363rd and 364th Fighter Squadrons.

Specifications

Crew: 2
Powerplant: 2 x 1276kW (1710hp) Rolls-Royce Merlin 12 cylinder piston engines
Maximum speed: 657km/h (408mph)
Range: 1963km (1220 miles)
Service ceiling: 9449m (31,000ft)
Dimensions: span 16.50m (54ft 2in); length 12.65m (41ft 6in); height 4.65m (15ft 3in)
Weight: 11,756kg (25,917lb) maximum takeoff
Armament: 4 x 20mm (0.8in) Hispano cannon and 4 x 7.7mm (0.303in) Browning MGs

▲ de Havilland Mosquito PR.Mk XVI (F-8)

653rd Bombardment Squadron / 25th Bombardment Group, Watton, mid-1944

Comprising the 652nd Bombardment Squadron (Heavy), 653rd Bombardment Squadron (Light) and 654th Bombardment Squadron (Special Purpose), the 25th Bombardment Group settled at Watton in April 1944. The heavy squadron flew the B-24, then B-17 four-engined bombers for weather reconnaissance over the Atlantic, while the light squadron flew the Mosquito in the same task over Europe.

Republic P-47D Thunderbolt

USAAF, European theatre, late 1944

Operating in a basic plain metal finish with only small areas of olive drab on the upper fuselage, the P-47D was used late in World War II primarily for the heavy fighter-bomber role, carrying considerable ordnance and exploiting its speed and sturdy airframe.

Specifications

Crew: 1
Powerplant: 1891kW (2535hp) Pratt & Whitney R-2800-59W Double Wasp
Maximum speed: 697km/h (433mph)
Range: 3060km (1900 miles) with drop tanks
Service ceiling: 12,495m (41,000ft)
Dimensions: span 12.42m (40ft 9in); length 11.02m (36ft 2in); height 4.47m (14ft 8in)
Weight: 7938kg (17,500lb) maximum take-off
Armament: 8 x 12.7mm (0.5in) MGs in wings, plus provision for 1134kg (2500lb) external bombs or rockets

US Eighth Air Force

1942–45

The Eighth Air Force was schemed as a major formation of the USAAF to be based in England. It would fly Boeing B-17 and Consolidated B-24 four-engined heavy bombers in precision attacks to destroy Germany's industrial capacity and its ability to make and sustain modern warfare.

The Eighth Air Force had the specific task of strategic heavy bombing, attacking pinpoint targets in German industrial, communications and resources facilities. This, the Americans believed, would materially damage the ability of the Germans to make and sustain a modern type of warfare.

Accuracy was the keynote of the concept, and was to be provided by the use of the advanced Norden bomb sight by heavy bombers flying by day, and at high altitude to reduce the chances of the German fighter arm being able to effect any significant interceptions. This bombing campaign lasted from the summer of 1942 to the end of the war in Europe in May 1945.

USAAF COMBAT GROUPS IN THE EUROPEAN THEATRE

Date	Light bomber	Medium bomber	Fighter
December 1942	1	6	14
December 1943	1	10	30
December 1944	4	13	29
March 1945	4	13	31

On 2 January 1942, the order creating the Eighth Air Force was signed, and the initial HQ was formed at Savannah, Georgia, on 28 January. On 8 January, it was announced that the organization of US Forces in the British Isles had been created, and the VIII

Bomber Command was established in England on 22 February 1942, with its initial HQ at RAF Bomber Command headquarters at High Wycombe. Additional commands of the Eighth Air Force were the VIII Air Support Command and VIII Fighter Command. During most of its existence in the UK, the Eighth Air Force had its headquarters at the Wycombe Abbey School for Girls in Buckinghamshire, and during World War II the Eighth Air Force was commanded successively by Major-General Carl A. Spaatz, Major General Ira C. Eaker, and Lieutenant-General 'Jimmy' Doolittle, the last best known as the leader of the April 1942 carrier-launched B-25 'Doolittle' raid on Tokyo and a few other Japanese targets. The Eighth Air Force later became the US Air Forces in Europe.

First blood

On 4 July 1942, six US crews from the 15th Bombardment Group (Light), together with six RAF crews, took off from RAF Swanton Morley, in Norfolk, on a daylight attack against four airfields in the German-occupied Netherlands. This was the first occasion on which US airmen had flown in

▲ North American P-51B Mustang

353rd Fighter Squadron / 354th Fighter Group, mid 1944

The P-51B was the first Mustang variant with the Packard V-1650 (Rolls-Royce Merlin) engine instead of the Allison V-1710 unit. This turned the Mustang into a superb fighter. Other features of this machine are the bulged 'Malcolm' hood for improved fields of vision, and a dorsal fin for enhanced directional authority.

Specifications

Crew: 1
Powerplant: 1044kW (1400hp) Packard V-1650 Rolls-Royce Merlin engine
Maximum speed: 690km/h (430mph)
Range: 3540km (2200 miles)
Service ceiling: 12,649m (41,500ft)
Dimensions: span 11.27m (37ft); length 9.84m (32ft 4in); height 4.15m (13ft 8in)
Weight: 4173kg (9200lb) loaded
Armament: 6 x 12.7mm (0.5in) MGs

▲ Supermarine Spitfire PR.Mk IX

14th Photographic Squadron / 7th Photographic Group, Mount Farm, 1944

Reaching its English base in July 1943, the 7th Photographic Group's components were the 13th, 14th, 22nd and 27th Photographic Squadrons. The group's standard aircraft early in its career were the F-4 and F-5 variants of the P-38 Lightning, but there were also some Spitfires and, later, the P-51D and P-51K.

Specifications

Crew: 1
Powerplant: 1170kW (1565hp) 12-cylinder Rolls-Royce Merlin 61 engine
Maximum speed: 642km/h (410mph)
Range: 698km (435 miles) on internal fuel tanks
Service ceiling: 12,650m (41,500ft)
Dimensions: span 11.23m (36ft 10in); length 9.47m (31ft 1in); height 3.86m (12ft 8in)
Weight: 3343kg (7370lb) loaded
Armament: 4 x 7.7mm (0.303in) MGs and 2 x 20mm (0.8in) cannons

American-built bombers against a German target. The event was of considerable historical importance, but in itself the raid was not in any way a significant success. Two of the US-crewed aircraft succumbed to German anti-aircraft fire. A start had been made, and before it ended its campaign the Eighth Air Force flew 332,904 sorties and dropped 633,260 tonnes (623,288 tons) of bombs on German targets on the European mainland. The campaign cost the Eighth Air Force large numbers of men and aircraft, losses peaking at 420 machines in April 1944.

USAAF TACTICAL WARPLANE STRENGTHS

Date	Light/medium bomber	Fighter
December 1941	1544	2170
December 1942	3757	5303
December 1943	6741	11,875
December 1944	9169	17,198
August 1945	8463	16,799

Specifications

Crew: 1
Powerplant: 1044kW (1400hp) Packard V-1650 Rolls-Royce Merlin engine
Maximum speed: 690km/h (430mph)
Range: 3540km (2200 miles)
Service ceiling: 12,649m (41,500ft)
Dimensions: span 11.27m (37ft); length 9.84m (32ft 4in); height 4.15m (13ft 8in)
Weight: 4173kg (9200lb) loaded
Armament: 6 x 12.7mm (0.5in) MGs

▲ North American P-51B Mustang

364th Fighter Squadron / 357th Fighter Group, Leiston, early 1944

Among the distinctions of the 357th Fighter Group were the facts that it was the first Mustang group in the Eighth Air Force, had a faster rate of aerial victories than any other Eighth Air Force group in the last 12 months of the war, and had the highest claims for one mission, on 14 January 1945, at 45 shot down and 56 probably downed.

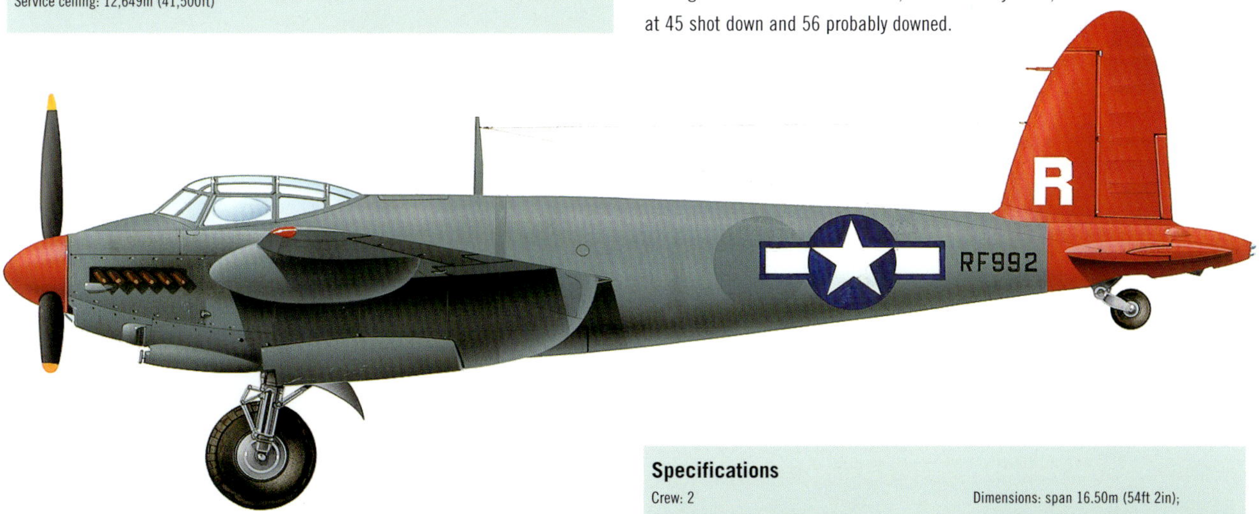

▲ de Havilland Mosquito PR.Mk XVI

25th Bombardment Group (Reconnaissance) / 325th Photographic Wing, Watton, summer 1944

Able to fly high and fast over German-held territory with only minimum chance of interception, the Mosquito PR.Mk XVI was used for weather reconnaissance.

Specifications

Crew: 2
Powerplant: 2 x 1276kW (1710hp) Rolls-Royce Merlin 12-cylinder piston engines
Maximum speed: 657km/h (408mph)
Range: 1963km (1220 miles)
Service ceiling: 9449m (31,000ft)
Dimensions: span 16.50m (54ft 2in); length 12.65m (41ft 6in); height 4.65m (15ft 3in)
Weight: 11,756kg (25,917lb) maximum takeoff
Armament: 4 x 20mm (0.8in) Hispano cannon and 4 x 7.7mm (0.303in) Browning MGs

1944 US Eighth Air Force Fighter Group

The standard fighter group of the Eighth Air Force in 1944 was three squadrons each of four flights, each flight being composed of two elements, each comprising a leader and his wing man. This organization gave the squadron 16 aircraft, and the group 48 aircraft. By this time, US production and training were running at rates so high that squadrons seldom lacked either aircraft or the pilots to man them.

Squadron 1

Flight 1 | **Flight 2** | **Flight 3** | **Flight 4**

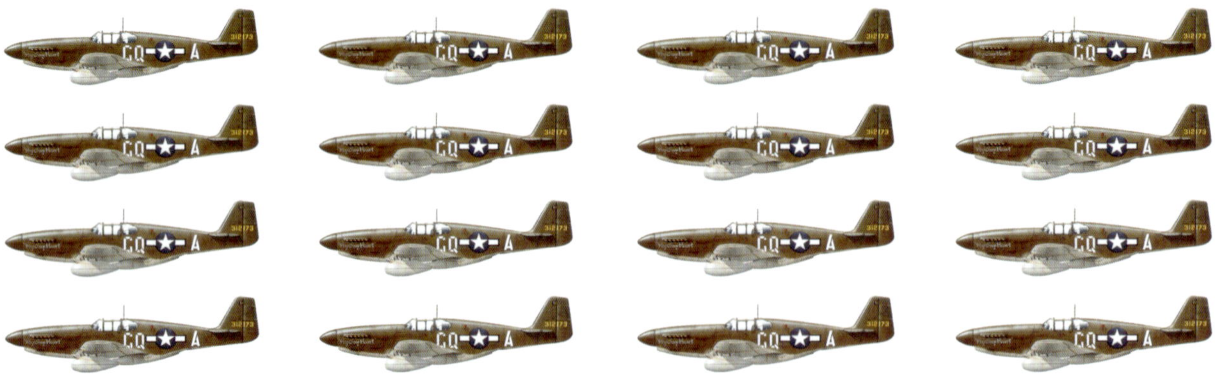

Squadron 2

Flight 1 | **Flight 2** | **Flight 3** | **Flight 4**

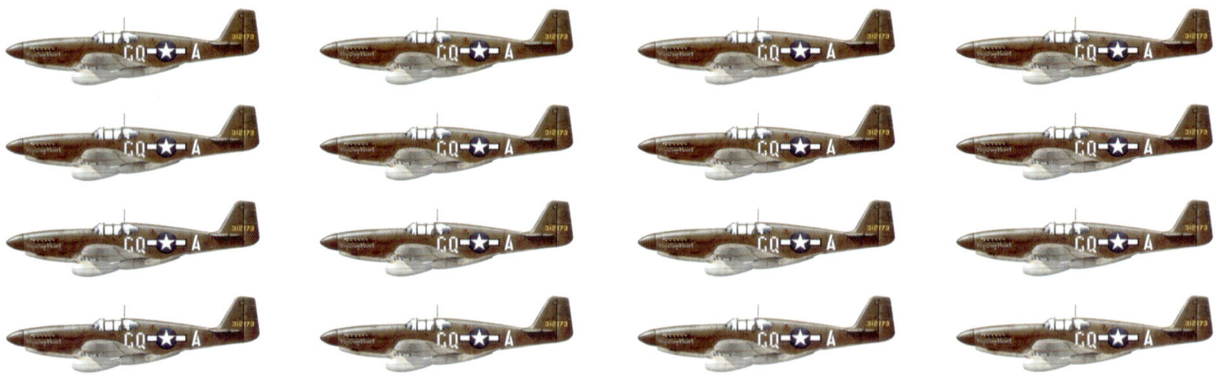

Squadron 3

Flight 1 | **Flight 2** | **Flight 3** | **Flight 4**

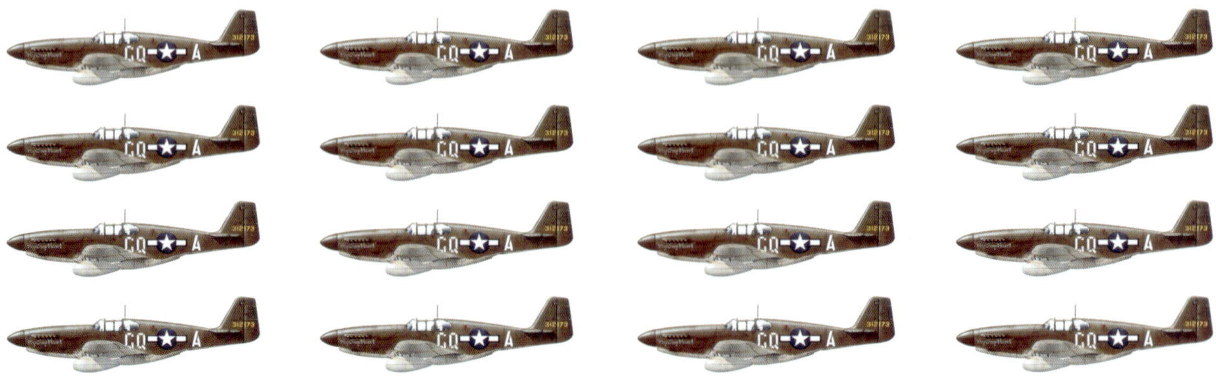

VIII Fighter Command

1942–45

VIII Fighter Command was created largely to provide escort for the bombers of VIII Bomber Command. Early tactics laying down the advisability of close escort gradually gave way to the practice of searching out the opposition and destroying it in the air and on the ground.

THE VIII FIGHTER Command was the fighter arm of the Eighth Air Force, and eventually consisted of 15 three-squadron groups in three wings. It was led for most of its existence by Major-General Willliam E. Kepner, who assumed command in August 1943. In June 1944, the command had six Thunderbolt, five Mustang and four Lightning groups, the last soon disappearing and only one unit still flying the Thunderbolt at the end of the war. Providing escort for VIII Bomber Command was the command's primary role.

ORGANIZATION

VIII Fighter Command
December 1943

- 65th Fighter Wing (five fighter groups)
- 66th Fighter Wing (five fighter groups)
- 67th Fighter Wing (five fighter groups)
- Three direct-reporting fighter groups

Limited range

The P-47 lacked the range to take bombers far beyond the German border, and the P-38 had high-altitude engine problems. But with the arrival of large numbers of P-51s early in 1944, the tide stared to turn. From January 1944, the fighters were no longer tied closely to the bombers but were freed to take the air war to the Germans at all altitudes, including their airfields. *Luftwaffe* losses rose to unsustainable levels, but bomber losses fell.

Specifications

Crew: 1
Powerplant: 1264kW (1695hp) Packard Merlin V-1650-7 V-12 piston engine
Maximum speed: 703km/h (437mph)
Range: 3347km (2080 miles)
Service ceiling: 12,770m (41,900ft)
Dimensions: span 11.28m (37ft); length 9.83m (32ft 3in); height 4.17m (13ft 8in)
Weight: 5488kg (12,100lb) loaded
Armament: 6 x 12.7mm (0.5in) MGs in wings, plus up to 2 x 454kg (1000lb) bombs

▲ **North American P-51D Mustang**

487th Fighter Squadron / 352nd Fighter Group, Chievres and Bodney, April 1945

This was the aeroplane of Major William Halton, commander of the squadron. By this late stage of the war, most US warplanes were flown in plain metal finish leavened by national, unit and personal markings.

Specifications

Crew: 1
Powerplant: 1264kW (1695hp) Packard Merlin V-1650-7 V-12 piston engine
Maximum speed: 703km/h (437mph)
Range: 3347km (2080 miles)
Service ceiling: 12,770m (41,900ft)
Dimensions: span 11.28m (37ft); length 9.83m (32ft 3in); height 4.17m (13ft 8in)
Weight: 5488kg (12,100lb) loaded
Armament: 6 x 12.7mm (0.5in) MGs in wings, plus up to 2 x 454kg (1000lb) bombs

▲ North American P-51D-5-NA Mustang

343rd Fighter Squadron / 55th Fighter Group, early 1945

This aircraft was flown by American ace Lt. E. Robert Welch. He notched up six kills and 12 aircraft destroyed on the ground. The red 'prancing horse' logo on the tailplane was used by the squadron from late 1944 onwards.

▲ North American P-51D Mustang

352nd Fighter Squadron / 353rd Fighter Group, late 1944–early 1945

Like most US fighter groups, the 353rd Fighter Group removed most of the camouflage from its P-51D and P-51K aircraft as being unnecessary in the face of the *Luftwaffe*'s decline, adding weight and increasing drag.

Specifications

Crew: 1
Powerplant: 1264kW (1695hp) Packard Merlin V-1650-7 V-12 piston engine
Maximum speed: 703km/h (437mph)
Range: 3347km (2080 miles)
Service ceiling: 12,770m (41,900ft)
Dimensions: span 11.28m (37ft); length 9.83m (32ft 3in); height 4.17m (13ft 8in)
Weight: 5488kg (12,100lb) loaded
Armament: 6 x 12.7mm (0.5in) MGs in wings, plus up to 2 x 454kg (1000lb) bombs

ALLIED FIGHTER PRODUCTION

Year	UK	USA
1939	1324	n/a
1940	4283	1162
1941	7064	4416
1942	9849	10,769
1943	10,727	23,988
1944	10,730	38,873
1945	5445	20,742

BOMB TONNAGES DROPPED IN EUROPE

Year	Bomber Command	US Eighth Air Force
1942	18,703	1411
1943	157,367	44,185
1944	525,518	389,119
1945	181,740	188,573
Totals	883,328	623,288

65th Fighter Wing
1942–45

While the fighter group, with its three squadrons, was the cornerstone of the Eighth Air Force's ability to provide tactical support for its bombers, it was administratively sensible to concentrate the groups into three high-level command bodies, namely the fighter wings.

The Eighth Air Force's fighter groups were assigned for administrative purposes to three fighter wings, each of them under the direct control of VIII Fighter Command until September 1945, when each of the three wings was assigned to one of the three bomb (later air) divisions. This restructing simplified the chain of command and also facilitated the planning of fighter support for the bombers of each division. It should be noted, however, that while each fighter wing was intended to support the bombers of its own division, in practice the fighter wings often provided support for the bombers of other divisions.

In the short term, VIII Fighter Command retained responsibility for three training organizations, namely the 495th and 496th Fighter Groups and the 1st Combat Crew Gunnery School, but late in November 1944 these too passed to the direct control

ORGANIZATION

65th Fighter Wing
1943–44

- 4th Fighter Group
- 56th Fighter Group
- 355th Fighter Group
- 361st Fighter Group
- 479th Fighter Group

ORGANIZATION

65th Fighter Wing
late 1944

- 4th Fighter Group (334th, 335th & 336th Fighter Squadrons)
- 56th Fighter Group (61st, 62nd & 63rd Fighter Squadrons)
- 355th Fighter Group (354th, 357th & 358th Fighter Squadrons)
- 361st Fighter Group (374th, 375th & 376th Fighter Squadrons)
- 479th Fighter Group (434th, 435th & 436th Fighter Squadrons)

Specifications

Crew: 1
Powerplant: 1264kW (1695hp) Packard Merlin V-1650-7 V-12 piston engine
Maximum speed: 703km/h (437mph)
Range: 3347km (2080 miles)
Service ceiling: 12,770m (41,900ft)
Dimensions: span 11.28m (37ft); length 9.83m (32ft 3in); height 4.17m (13ft 8in)
Weight: 5488kg (12,100lb) loaded
Armament: 6 x 12.7mm (0.5in) MGs in wings, plus up to 2 x 454kg (1000lb) bombs

▲ **North American P-51D Mustang**

375th Fighter Squadron / 361st Fighter Group, early 1945

The P-51 was a superb example of the fighter designer's art, with excellently harmonized controls. The ventral radiator installation lowered drag and actually produced some thrust as the heated cooling air was discharged, and the contents of the cheap impregnated paper drop tanks added considerable range.

▲ **'Razorback' design**

These early P-47Bs have the classic 'razorback' design. The lead aircraft is flown by P-47 ace Hubert Zemke.

of the air divisions. The VIII Fighter Command order of battle was based on three fighter wings, as noted above, and the lowest numbered of these was the 65th Fighter Wing, which had its headquarters at the Dane Bradbury School at Saffron Walden, and in September 1944 was reassigned from the VIII Fighter Command to the 2nd Bomb Division (from January 1945, the 2nd Air Division).

The 65th Fighter Wing's components consisted of the following: the 4th Fighter Group 'The Eagles' (the 334th, 335th and 336th Fighters Squadrons with the Spitfire, Thunderbolt and Mustang respectively) based at Debden from September 1942 to April 1943, and then at Steeple Morden to the end of the war; the 56th Fighter Group (61st, 62nd and 63rd Fighter Squadrons with the Thunderbolt) based at King's Cliffe, Horsham St Faith, Halesworth, Boxted and Little Walden between January 1943 and the end of the war; the 355th Fighter Group (354th, 357th and

Specifications

Crew: 1

Powerplant: 1264kW (1695hp) Packard Merlin V-1650-7 V-12 piston engine

Maximum speed: 703km/h (437mph)

Range: 3347km (2080 miles)

Service ceiling: 12,770m (41,900ft)

Dimensions: span 11.28m (37ft); length 9.83m (32ft 3in); height 4.17m (13ft 8in)

Weight: 5488kg (12,100lb) loaded

Armament: 6 x 12.7mm (0.5in) MGs in wings, plus up to 2 x 454kg (1000lb) bombs

▲ **North American P-51D Mustang**

355th Fighter Group, Steeple Morden, 1945

This was the aeroplane of the group's commander, Lieutenant-Colonel Claiborne H. Kinnard, Jr., who led the group between 21 February and 7 June 1945.

358th Fighter Squadrons with Thunderbolt and Mustang) based at Steeple Morden from July 1943 to the end of the war; the 361st Fighter Group (374th, 375th and 376th Fighter Squadrons with Thunderbolt and Mustang) based at Bottisham, Little Walden, St Dizier, Chievres and, finally, Little Walden once again; and lastly the 479th Fighter Group, known as 'Riddle's Raiders' (434th, 435th and 436th Fighter Squadrons with the Lightning and Mustang), based at Wattisham from May 1944.

THUNDERBOLT ACES OF THE 56TH FIGHTER GROUP

Name	Unit	Victories
Francis Gebreski	56th FG	28
Robert S. Johnson	56th FG	27
David Schilling	56th FG	22.5
Fred Christenson	56th FG	21.5
Walter Mahurin	56th FG	19
Hubert Zemke	56th FG	18
Gerald Johnson	56th FG	16.5

66th Fighter Wing

1942–45

Like the other fighter wings of the VIII Fighter Command, the 66th Fighter Wing was based on the standard five-group organization, and from its inception in December 1942 became one of the finest fighting elements of the Eighth Air Force.

The 66th Fighter Wing had its headquarters at Sawston Hall, near Cambridge, and was later transferred from the control of the VIII Fighter Command to the 3rd Bomb Division (later 3rd Air Wing). The 66th Fighter Wing was of standard fighter wing strength inasmuch as it controlled five fighter groups.

ORGANIZATION

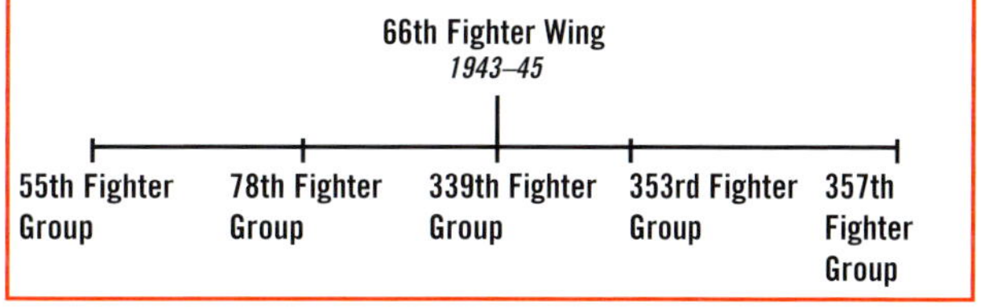

Specifications

Crew: 1
Powerplant: 1264kW (1695hp) Packard Merlin V-1650-7 V-12 piston engine
Maximum speed: 703km/h (437mph)
Range: 3347km (2080 miles)
Service ceiling: 12,770m (41,900ft)
Dimensions: span 11.28m (37ft); length 9.83m (32ft 3in); height 4.17m (13ft 8in)
Weight: 5488kg (12,100lb) loaded
Armament: 6 x 12.7mm (0.5in) MGs in wings, plus up to 2 x 454kg (1000lb) bombs

▲ North American P-51K Mustang

362nd Fighter Squadron / 357th Fighter Group, Leiston, early 1945

From December 1944 many of the 357th Fighter Group's P-51D and P-51K fighters were stripped of camouflage, but for anti-glare purposes retained green paint on the upper fuselage forward of the cockpit.

The 55th Fighter Group was based at Nuthampstead from September 1943 and then Wormingford from April 1944 to the end of the war. It had the 38th, 338th and 343rd Fighter Squadrons flying Lightnings and then Mustangs. The 78th Fighter Group was based at Goxhill from December 1942 and then Duxford from April 1943 to the end of the war, and had the 82nd, 83rd and 84th Fighter Squadrons with the Lightning, Thunderbolt and Mustang. The 339th Fighter Group based at Fowlmere from April 1944 to the end of the war had the 503rd, 504th and 505th Fighter Squadrons with the Mustang. The 353rd Fighter Group was based at Goxhill between June and August 1943, Metfield between August 1943 and April 1944, and finally Raydon for the rest of the war. It had the 350th, 351st and 352nd Fighter Squadrons with the Thunderbolt and then the Mustang. Finally the 357th Fighter Group, based at Raydon between November 1943 and January 1944, and then Leiston for the rest of the war, had the 362nd, 363rd and 364th Fighter Squadrons with the Mustang, which it introduced in the Eighth Air Force.

▲ **North American P-51K Mustang**

362nd Fighter Squadron / 357th Fighter Group, Leiston, late 1944

This was the aeroplane of Captain Leonard Kit Carson, who on 27 November 1944 shot down five German fighters in a single engagement.

Specifications

Crew: 1
Powerplant: 1264kW (1695hp) Packard Merlin V-1650-7 V-12 piston engine
Maximum speed: 703km/h (437mph)
Range: 3347km (2080 miles)
Service ceiling: 12,770m (41,900ft)
Dimensions: span 11.28m (37ft); length 9.83m (32ft 3in); height 4.17m (13ft 8in)
Weight: 5488kg (12,100lb) loaded
Armament: 6 x 12.7mm (0.5in) MGs in wings, plus up to 2 x 454kg (1000lb) bombs

Specifications

Crew: 1
Powerplant: 1264kW (1695hp) Packard Merlin V-1650-7 V-12 piston engine
Maximum speed: 703km/h (437mph)
Range: 3347km (2080 miles)
Service ceiling: 12,770m (41,900ft)
Dimensions: span 11.28m (37ft); length 9.83m (32ft 3in); height 4.17m (13ft 8in)
Weight: 5488kg (12,100lb) loaded
Armament: 6 x 12.7mm (0.5in) MGs in wings, plus up to 2 x 454kg (1000lb) bombs

▲ **North American P-51D Mustang**

362nd Fighter Squadron / 357th Fighter Group, late 1944

The omission of most camouflage paint, which was not needed in the absence of effective German fighter opposition, saved maintenance time and cost, and added slightly to the aeroplane's maximum speed.

67th Fighter Wing

1943–45

Another five-group formation, the 67th Fighter Wing provided escort for the bombers of the Eighth Air Force, and was then let off the leash to a modest degree to serve the bombers better by taking the war to the German fighter arm on the ground as well as in the air.

THE 67TH FIGHTER Wing was headquartered at Walcot Hall, near Stamford, and in September 1944 was reallocated from the VIII Fighter Command to the 1st Bomb Division (from January 1945 the 1st Air Division, an integrated formation with its own bomber and fighter elements). The 67th Fighter Wing had the standard five fighter groups.

The 20th Fighter Group was based at King's Cliffe from August 1943, and its three component units were the 55th, 77th and 79th Fighter Squadrons, which flew the P-38 Lightning to July 1944, when they transitioned to the P-51 Mustang for the rest of the war. The 352nd Fighter Group was based at Bodney from July 1943, with detachments at Asche and Chievres on the European mainland between December 1944 and April 1945. Its components were the 328th, 486th and 487th Fighter Squadrons with the Thunderbolt in its P-47D variant between July 1943 and April 1944, and then the Mustang in its P-51B, P-51C, P-51D and P-51K variants thereafter for the rest of the war. The 356th Fighter Group was based at Goxhill between August and October 1943 before settling at Martlesham Heath for the rest of the war, and its components were the 359th, 360th and 361st Fighter Squadrons, which flew the Thunderbolt in its P-47D form to November 1944 before transitioning to the Mustang in its P-51D and P-51K forms for the rest of the war.

The 359th Fighter Group was based at East Wretham from October 1943 to the end of the war, and its components were the 368th, 369th and 370th Fighter Squadrons, which flew the Thunderbolt in its P-47D form to May 1944, when they transitioned to the Mustang, in its P-51B, P-51C, P-51D and P-51K forms, for the rest of the war.

The 364th Fighter Group was based at Honington from February 1944 to the end of the war, and its components were the 382rd, 384th and 385th Fighter Squadrons, which flew the Lightning in its P-38J form to July 1944 before converting to the Mustang, in its P-51D form, for the rest of the war.

ORGANIZATION

67th Fighter Wing
1943–45

- 20th Fighter Group
- 352nd Fighter Group
- 356th Fighter Group
- 359th Fighter Group
- 364th Fighter Group

P-51 power

This P-51D Mustang carries the auxiliary fuel tanks introduced in late 1943 for long-range bomber escort duties.

MUSTANG ACES OF THE EIGHTH AIR FORCE

Name	Unit	Victories
George Preddy	352nd FG	25.83
John Meyer	352nd FG	24
Ray Wetmore	359th FG	21.25
Don Gentile	4th FG	19.83
Duncan Glenn	353rd FG	19
Walter Beckham	353rd FG	18
Duane Beeson	4th FG	17.33

Specifications

Crew: 1
Powerplant: 1891kW (2535hp) Pratt & Whitney R-2800-59W Double Wasp
Maximum speed: 697km/h (433mph)
Range: 3060km (1900 miles) with drop tanks
Service ceiling: 12,495m (41,000ft)
Dimensions: span 12.42m (40ft 9in); length 11.02m (36ft 2in); height 4.47m (14ft 8in)
Weight: 7938kg (17,500lb) maximum take-off
Armament: 8 x 12.7mm (0.5in) MGs in wings, plus provision for 1134kg (2500lb) external bombs or rockets

▲ Republic P-47D Thunderbolt

352nd Fighter Squadron / 353rd Fighter Group, Raydon, July 1944

The Thunderbolt was much improved as a fighter in its later forms with the original pattern of framed canopy, forward of the 'razorback' upper rear fuselage, replaced by a clear-view canopy and cut-down rear decking.

Specifications

Crew: 1
Powerplant: 1264kW (1695hp) Packard Merlin V-1650-7 V-12 piston engine
Maximum speed: 703km/h (437mph)
Range: 3347km (2080 miles)
Service ceiling: 12,770m (41,900ft)
Dimensions: span 11.28m (37ft); length 9.83m (32ft 3in); height 4.17m (13ft 8in)
Weight: 5488kg (12,100lb) loaded
Armament: 6 x 12.7mm (0.5in) MGs in wings, plus up to 2 x 454kg (1000lb) bombs

▲ North American P-51D Mustang

369th Fighter Squadron / 359th Fighter Group, East Wretham, November 1944

This was the aeroplane of 1st Lieutenant Claude Crenshaw, who ended the war with seven air and three ground 'kills'. His best day was 21 November 1944, when Crenshaw downed four Focke-Wulf Fw 190 fighters.

Specifications

Crew: 1
Powerplant: 1170kW (1565hp) 12-cylinder Rolls-Royce Merlin 61 engine
Maximum speed: 642km/h (410mph)
Range: 698km (435 miles) on internal fuel tanks
Service ceiling: 12,650m (41,500ft)
Dimensions: span 11.23m (36ft 10in); length 9.47m (31ft 1in); height 3.86m (12ft 8in)
Weight: 3343kg (7370lb) loaded
Armament: 4 x 7.7mm (0.303in) MGs and 2 x 20mm (0.8in) cannons

▲ Supermarine Spitfire PR.Mk IX

7th Photographic Group

The group flew a combination of F-5 (P-38), P-51 and Spitfire IX photo/recon aircraft to obtain information about bombardment targets and damage inflicted by bombardment operations. The group also provided mapping service for air and ground units; observed and reported on enemy transportation, installations, and positions; and obtained data on weather conditions.

▲ North American P-51B Mustang

354th Fighter Squadron / 355th Fighter Group, Steeple Morden, April 1944

This was the aeroplane, in standard olive drab and neutral grey colour, of 2nd Lieutenant Henry Brown, who ended the war with 17.2 kills in air-to-air combat.

Specifications

Crew: 1
Powerplant: 1044kW (1400hp) V-1650-3 engine
Maximum speed: 690km/h (430mph)
Range: 3540km (2200 miles)
Service ceiling: 12,649m (41,500ft)
Dimensions: span 11.27m (37ft); length 9.84m (32ft 4in); height 4.15m (13ft 8in)
Weight: 4173kg (9200lb) loaded
Armament: 6 x 12.7mm (0.5in) MGs

▲ North American P-51B Mustang

487th Fighter Squadron / 352nd Fighter Group, Bodney, May 1944

This was the aeroplane of 1st Lieutenant William Wisner, who achieved 16 victories in this and one other aeroplane.

Specifications

Crew: 1
Powerplant: 1044kW (1400hp) V-1650-3 engine
Maximum speed: 690km/h (430mph)
Range: 3540km (2200 miles)
Service ceiling: 12,649m (41,500ft)
Dimensions: span 11.27m (37ft); length 9.84m (32ft 4in); height 4.15m (13ft 8in)
Weight: 4173kg (9200lb) loaded
Armament: 6 x 12.7mm (0.5in) MGs

▲ North American P-51B Mustang

336th Fighter Squadron / 4th Fighter Group, Debden, May 1944

This was the aeroplane of Captain Willard 'Millie' Milikan, who achieved 13 victories despite earlier rejection by the US Army Air Corps, and a poor rating for flying skill when he joined the Royal Canadian Air Force. Milikan commanded a unit equipped with Republic F-84 jet-powered warplanes in the Korean War.

Specifications

Crew: 1
Powerplant: 1081kW (1450hp) V-1650-3 engine
Maximum speed: 690km/h (430mph)
Range: 1215km (755 miles)
Service ceiling: 12,649m (41,500ft)
Dimensions: span 11.27m (37ft); length 9.84m (32ft 4in); height 4.15m (13ft 8in)
Weight: 4173kg (9200lb) loaded
Armament: 4 x 12.7mm (0.5in) MGs

Specifications

Crew: 1
Powerplant: 1044kW (1400hp) Packard V-1650-3 Rolls-Royce Merlin engine
Maximum speed: 690km/h (430mph)
Range: 3540km (2200 miles)
Service ceiling: 12,649m (41,500ft)
Dimensions: span 11.27m (37ft); length 9.84m (32ft 4in); height 4.15m (13ft 8in)
Weight: 4173kg (9200lb) loaded
Armament: 6 x 12.7mm (0.5in) MGs

▲ North American P-51B Mustang

328th Fighter Squadron / 352nd Fighter Group, Bodney, July 1944

Fitted with a 'Malcolm hood', this P-51B was the aeroplane of 1st Lieutenant John F. Thornell, Jr., one of the 352nd Fighter Group's leading aces with 17.25 victories.

Specifications

Crew: 1
Powerplant: 1264kW (1695hp) Packard Merlin V-1650-7 V-12 piston engine
Maximum speed: 703km/h (437mph)
Range: 3347km (2080 miles)
Service ceiling: 12,770m (41,900ft)
Dimensions: span 11.28m (37ft); length 9.83m (32ft 3in); height 4.17m (13ft 8in)
Weight: 5488kg (12,100lb) loaded
Armament: 6 x 12.7mm (0.5in) MGs in wings, plus up to 2 x 454kg (1000lb) bombs

▲ North American P-51D Mustang

354th Fighter Squadron / 355th Fighter Group, Steeple Morden, late summer 1944

This was the second aircraft of Captain Henry Brown, leading ace of the 355th Fighter Group (14.2 air-to-air and 14.5 air-to-surface victories) before being taken prisoner on 3 October 1944.

Specifications

Crew: 1
Powerplant: 1264kW (1695hp) Packard Merlin V-1650-7 V-12 piston engine
Maximum speed: 703km/h (437mph)
Range: 3347km (2080 miles)
Service ceiling: 12,770m (41,900ft)
Dimensions: span 11.28m (37ft); length 9.83m (32ft 3in); height 4.17m (13ft 8in)
Weight: 5488kg (12,100lb) loaded
Armament: 6 x 12.7mm (0.5in) MGs in wings, plus up to 2 x 454kg (1000lb) bombs

▲ North American P-51D Mustang

343rd Fighter Squadron / 55th Fighter Group, Worningford, September 1944

This was the aeroplane of Lieutenant E. Robert Welch, who achieved 12 victories.

56th Fighter Group – 1944

Based at Boxted from 18 April 1944, and commanded in this period by Colonel Hubert A. Zemke until 12 August, then by Colonel David C. Schilling until 27 January 1945, and finally (in World War II) by Lieutenant-Colonel Lucian A. Dade, Jr., the 56th Fighter Group was of the standard triangular pattern of the USAAF. Its three squadrons, each comprising four flights of four aircraft (two leader and wingman pairs) for 48 aircraft in all, were the 61st, 62nd and 63rd Fighter Squadrons.

61st Fighter Squadron

Flight 1 **Flight 2** **Flight 3** **Flight 4**

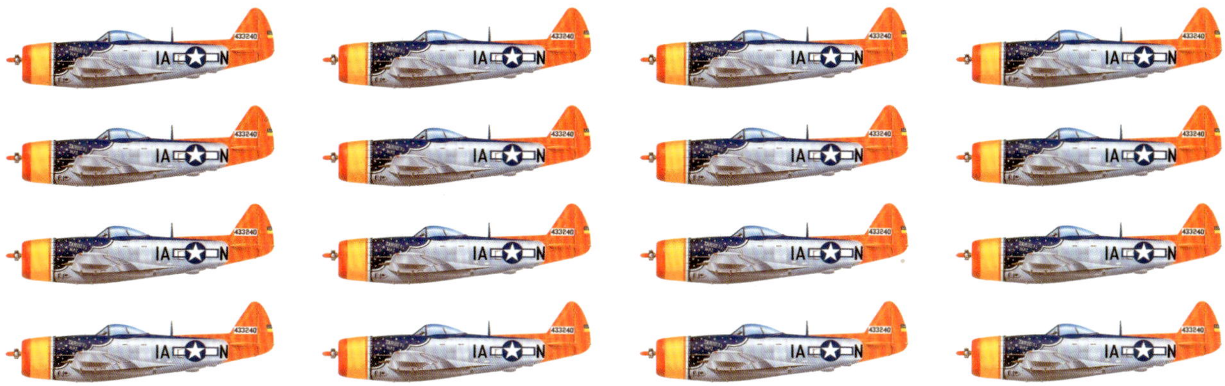

62nd Fighter Squadron

Flight 1 **Flight 2** **Flight 3** **Flight 4**

63rd Fighter Squadron

Flight 1 **Flight 2** **Flight 3** **Flight 4**

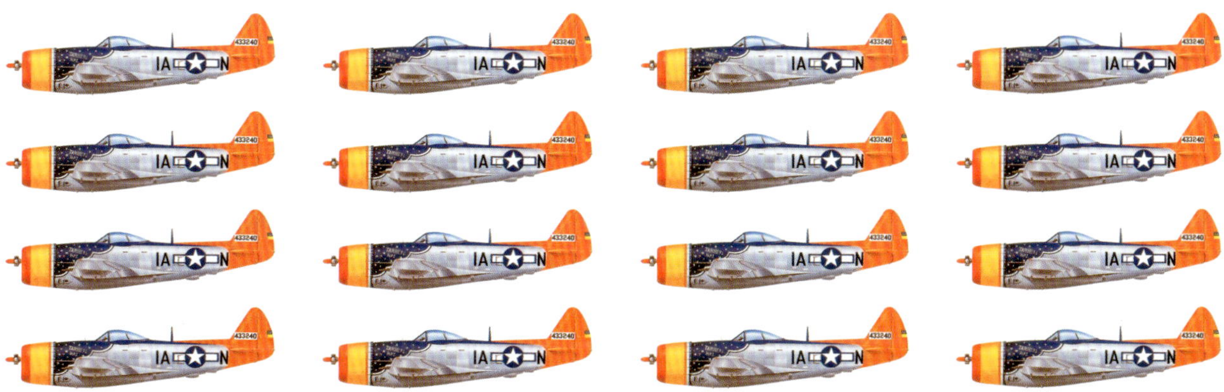

US Ninth Air Force

1942–45

Created for service in North Africa as a multi-role air force, the Ninth Air Force came into its own when translated into a highly capable tactical air force to support the US armies in the northwest Europe campaign of 1944–45.

THE US NINTH AIR FORCE was the tactical counterpart of the Eighth Air Force, initially intended for service in North Africa, and was created on 12 November 1942 to win air superiority, prevent Axis forces from supplying or rebuilding their forces, and provide the ground forces with close support.

By the end of 1942, a total of 370 aircraft had been ferried to the Ninth Air Force, the majority of them fighter-bombers, medium bombers and heavy bombers. In February 1943, the Germans took the offensive and drove through the Kasserine Pass in southern Tunisia. The US II Corps took a beating, but the Germans were checked with the aid of Ninth and Twelfth Air Force elements.

US NINTH ARMY AIR FORCE (1942–43)

Unit	Aircraft type
IX Fighter Command	
57th FG	P-40F
79th FG	P-40F
324th FG	P-40F
IX Bomber Command	B-24D/B-25C
316th Troop Carrier Group	C-47

ORGANIZATION

Ninth Air Force
December 1943

- IX Bomber Command (four bombardment groups)
- IX Fighter Command (three fighter groups)
- 316th Troop Carrier Group

Sicilian campaign

The Axis forces in North Africa surrendered in May 1943, and the Allies prepared to invade Sicily. After the Allied victory in Tunisia, Ninth Air Force groups attacked airfields and rail facilities in Sicily and Italy, and then in July supported the Allied forces in the Sicilian campaign.

Specifications

Crew: 2

Powerplant: 2 x 1194kW (1600hp) Allison V-1710-111/113

Maximum speed: 666km/h (414mph)

Range: 4184km (2600 miles)

Service ceiling: 13,410m (20,000ft)

Dimensions: span 15.85m (52ft); length 11.53m (37ft 10in); height 2.99m (9ft 10in)

Weight: 9798kg (21,600lb) maximum takeoff

Armament: 1 x 20mm (0.8in) cannon and 4 x 12.7mm (0.5in) MGs in nose, plus 2 x 907kg (2000lb) bombs

▲ **Lockheed P-38J Lightning**

401st Fighter Squadron / 370th Fighter Group, Florennes, Belgium, November 1944

By this late stage of World War II, the Lightning was used mostly in the long-range fighter-bomber role, a low-altitude task in which the altitude limitations of its Allison V-1710 liquid-cooled V-12 engines were not a problem.

Specifications

Crew: 2
Powerplant: 2 x 1194kW (1600hp) Allison V-1710-111/113
Maximum speed: 666km/h (414mph)
Range: 4184km (2600 miles)
Service ceiling: 13,410m (20,000ft)
Dimensions: span 15.85m (52ft); length 11.53m (37ft 10in); height 2.99m (9ft 10in)
Weight: 9798kg (21,600lb) maximum takeoff
Armament: 1 x 20mm (0.8in) cannon and 4 x 12.7mm (0.5in) MGs in nose, plus 2 x 907kg (2000lb) bombs

▲ Lockheed P-38J Lightning

79th Fighter Squadron / 20th Fighter Group, early 1944

Its size and weight militated against use of the twin-engined Lighting as an air combat fighter, but its high performance, especially in speed and range, suggested that the type might find gainful employment in the bomber escort role until it found its métier in the long-range fighter-bomber and attack roles.

Specifications

Crew: 2
Powerplant: 2 x 1194kW (1600hp) Allison V-1710-111/113
Maximum speed: 666km/h (414mph)
Range: 4184km (2600 miles)
Service ceiling: 13,410m (20,000ft)
Dimensions: span 15.85m (52ft); length 11.53m (37ft 10in); height 2.99m (9ft 10in)
Weight: 9798kg (21,600lb) maximum takeoff
Armament: 1 x 20mm (0.8in) cannon and 4 x 12.7mm (0.5in) MGs in nose, plus 2 x 907kg (2000lb) bombs

▲ Lockheed P-38 Lightning

365th Fighter Squadron / 358th Fighter Group, England, mid-1944

On 1 February 1944, the Eighth Air Force's 358th Fighter Group was passed to the Ninth Air Force, which in turn reallocated its Mustang-equipped 357th Fighter Group to the Eighth Air Force.

IX FIGHTER COMMAND (NOVEMBER 1943)

Unit	Base	Aircraft type
67th Recon Group	RAF Membury	F-5/P-38
354th FG	RAF Boxted	P-51
357th FG (to 1 Feb 44)	RAF Leiston	P-51
358th FG (from 1 Feb 44)	RAF Leiston	P-47

In August and September 1943, the Ninth Air Force was tasked to move to the UK, and transferred its units in North Africa to the Twelfth Air Force. The Ninth Air Force was deactivated in Egypt on 16 October 1943, the day on which its new HQ was reactivated at Burtonwood in England, and then became a decisive tactical air force led by Lieutenant-General Hoyt S. Vandenberg.

The nucleus of the Ninth Air Force was formed in November 1943 by the transfer of some Eighth Air Force tactical bomber, fighter and troop carrier groups, and during the following winter the Ninth Air Force expanded very rapidly so that by the end of May 1944, when the last combat group became operational, its complement ran to 45 flying groups operating some 5000 aircraft.

Together with the Eighth Air Force, the Ninth Air Force was tasked with destroying the *Luftwaffe* in the air and on the ground and thus gaining air supremacy before the launch of the Allied invasion of France in June 1944. By early August, most Ninth Air Force groups had been transferred to bases in France, and were then assigned to tactical air commands (TACs) supporting the ground forces: the XXIX TAC supported the Ninth Army in the north, the IX TAC the First Army in the centre, and the XIX TAC the Third Army in the south; air cover was provided by the IX Air Defense Command.

In December 1944 and January 1945, the Ninth Air Force's fighters and bombers were critical in defeating the Germans in the so-called Battle of the Bulge, and for the rest of the war the Ninth Air Force provided excellent support for the US forces. With victory won, the Ninth Air Force was deactivated on 2 December 1945.

Specifications

Crew: 1
Powerplant: 1044kW (1400hp) Packard V-1650-3 Rolls-Royce Merlin engine
Maximum speed: 690km/h (430mph)
Range: 3540km (2200 miles)
Service ceiling: 12,649m (41,500ft)
Dimensions: span 11.27m (37ft); length 9.84m (32ft 4in); height 4.15m (13ft 8in)
Weight: 4173kg (9200lb) loaded
Armament: 6 x 12.7mm (0.5in) MGs

▲ **North American P-51B Mustang**

355th Fighter Squadron / 354th Fighter Group, England, late 1943

This is an example of the first Mustang variant with the Packard V-1650 (Rolls-Royce Merlin) liquid-cooled V-12 engine, and is carrying 284l (75 US gal) drop tanks for extended range.

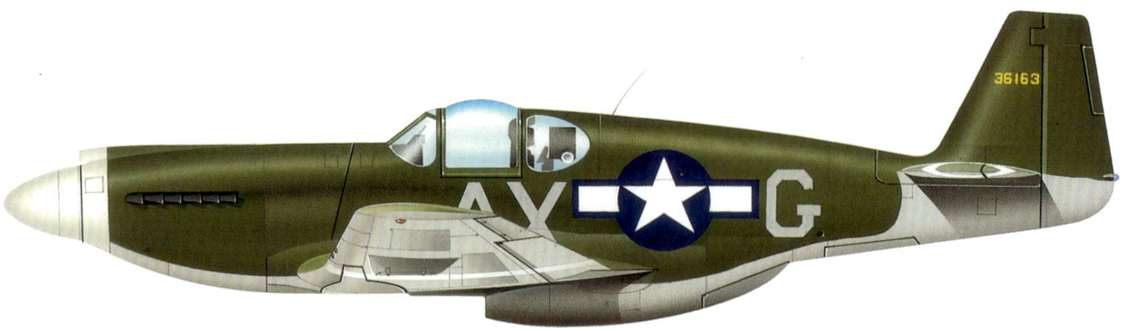

Specifications

Crew: 1
Powerplant: 1044kW (1400hp) Packard V-1650-3 Rolls-Royce Merlin engine
Maximum speed: 690km/h (430mph)
Range: 3540km (2200 miles)
Service ceiling: 12,649m (41,500ft)
Dimensions: span 11.27m (37ft); length 9.84m (32ft 4in); height 4.15m (13ft 8in)
Weight: 4173kg (9200lb) loaded
Armament: 6 x 12.7mm (0.5in) MGs

▲ **North American F-6B Mustang**

107th Tactical Reconnaissance Squadron, Europe, 1944

The F-6B was the tactical reconnaissance derivative of the P-51A fighter, with the basic armament retained but a pair of K-24 cameras installed.

Normandy and Northwest Europe

June 1944 – May 1945

The Ninth Air Force offered the armies of the US 12th Army Group the type of tactical air support that the British 2nd Tactical Air Force gave to the Anglo-Canadian 21st Army Group in the northwest Europe campaign, greatly speeding the Allied advance and reducing losses.

The Ninth Air Force was a key element in the planning for the debouchment of the US forces in *Overlord*, the Allied amphibious invasion of France over the beaches of Normandy in June 1944. The Ninth Air Force had already played a major role in destroying the German lines of communication in France and the Low Countries behind all the possible invasion areas, and thus isolating the German forces and depriving them of a realistic chance of reinforcement or major resupply.

Like the British 2nd TAF, the US Ninth Air Force had also been instrumental with the Allied heavy bomber forces in destroying much of the coastal defence system that the Germans had built so expensively and laboriously over the previous years, with particular emphasis placed on the destruction of the artillery emplacements that could otherwise have crippled the seaborne approach of the Allied invasion forces.

ORGANIZATION

Allied Expeditionary Air Force
June 1944

- British 2nd Tactical Air Force
- US Ninth Air Force
- Allied Strategic Air Force (British Bomber Command & US Eighth Air Force)

US FIGHTER-BOMBER ARMAMENT

Type	Guns	Disposable
Bell P-39Q	1 x 20mm (0.79in) 4 x 12.7mm (0.5in)	227kg (500 lb)
Curtiss P-40N	6 x 12.7mm (0.5in)	680kg (1500lb)
Lockheed P-38L	1 x 20mm (0.79in) 4 x 12.7mm (0.5in)	1814kg (4000lb)
North American P-51D	6 x 12.7mm (0.5in)	907kg (2000lb)
Republic P-47D	8 x 12.7mm (0.5in)	1134kg (2500lb)

Specifications

Crew: 1
Powerplant: 1891kW (2535hp) Pratt & Whitney R-2800-59W Double Wasp
Maximum speed: 697km/h (433mph)
Range: 3060km (1900 miles) with drop tanks
Service ceiling: 12,495m (41,000ft)
Dimensions: span 12.42m (40ft 9in); length 11.02m (36ft 2in); height 4.47m (14ft 8in)
Weight: 7938kg (17,500lb) maximum takeoff
Armament: 8 x 12.7mm (0.5in) MGs in wings, plus provision for 1134kg (2500lb) external bombs or rockets

▲ **Republic P-47D Thunderbolt**

336th Fighter Squadron / 358th Fighter Group, Toul, France, late 1944

Later production examples of the P-47D introduced a small dorsal fillet to reduce a tail flutter problem. This aeroplane carries the orange tail markings of the 1st Tactical Air Force.

With the ground forces ashore and the lodgement area enlarged steadily if slowly, the fighter and then the medium bomber groups of the Ninth Air Force could be transferred from England to France, either extending their tactical radii deeper into German-held territory or increasing their loiter times in the areas over or immediately beyond the frontline.

Hard fighting

Thereafter, the Ninth Air Force moved east through France and Belgium toward the western part of Germany. There was occasion for hard fighting in the air, but as often as not it was the weather rather than the Germans that was the greater hindrance. Air support was provided for the ground forces with concentration and accuracy, and in the later stages of the campaign this proved of immense value in speeding the US advance and reducing losses. During the invasion of southern France in August 1944, two fighter groups of the Ninth Air Force were reallocated to the provisional US/French 1st TAF supporting the invasion force's drive north, and for the Arnhem operation of September the Ninth Air Force transferred its IX Troop Carrier Command (14 C-47 groups) to the 1st Allied Airborne Army.

In December 1944 and January 1945, the Ninth Air Force committed its warplanes, as and when the weather permitted, to stem and then defeat the German offensive in the 'Battle of the Bulge'. The Ninth Air Force's troop carrier groups delivered paratrooper and glider units during the Allied airborne crossing of the Rhine on 24 March 1945, which was the largest single airborne drop in history.

Photo reconnaisance

An F-4 Lightning of the 3rd Photographic Reconnaissance Group banks towards the photographer. The 3rd Group arrived in Algeria in late 1942 and flew countless missions in support of the Allied armies in Europe.

IX Fighter Command

June 1944 – May 1945

The IX Fighter Command was one of the three formations in the US Ninth Air Force and, despite its designation, it was optimized for the fighter-bomber role in support of the ground forces. The command did have a number of Mustang units, however, offering some genuine fighter ability.

In the summer of 1944, command of the US Ninth Air Force was in the hands of Lieutenant-General Lewis H. Brereton, an officer with widespread command experience in the Far East and Middle East before he arrivd in the UK. On 24 August, Brereton was reassigned to become the commander of the Allied First Airborne Army, and his place at the head of the US Ninth Air Force was subsequently assumed by Lieutenant-General Hoyt S. Vandenberg. IX Fighter Command was headquartered at Middle Wallop in the south of England until it moved to Les Obeaux in France during August and then five other towns in France, Belgium and Germany as the US forces moved east. It was commanded by Major-General R. Etheral, and from the summer of 1944 was largely an administrative organization that ensured the smooth operation of the eventual total of the three tactical air commands for which it had responsibility. These were the IX, XIX and XXIX TACs, allocated to support the initial trio of US armies that took the ground war to the Germans from 6 June 1944.

ORGANIZATION

IX Fighter Command
late 1944

IX Tactical Air Command	XIX Tactical Air Command	XXIX Tactical Air Command

FIGHTER ACES OF THE NINTH AIR FORCE

Name	Victories
Glenn T. Eagleston	18.5
Don M. Beerbower	15.5
Jack T. Bradley	15
Bruce W. Carr	14
Wallace N. Emmer	14
Kenneth H. Dahlberg	14
Robert W. Stephens	13
Lowell K. Brueland	12.5
James H. Howard	12.3
Clyde B. East	12

North American P-51D Mustang

354th Fighter Group, European theatre, 1944–45

As suggested by the motif pained on the cowling of the Packard V-1650 liquid-cooled V-12 engine, this is the aeroplane of Lieutenant-Colonel Glenn T. Eagleston, commander of the 354th Fighter Group and an ace with 24 confirmed victories.

Specifications

Crew: 1
Powerplant: 1264kW (1695hp) Packard Merlin V-1650-7 V-12 piston engine
Maximum speed: 703km/h (437mph)
Range: 3347km (2080 miles)
Service ceiling: 12,770m (41,900ft)
Dimensions: span 11.28m (37ft); length 9.83m (32ft 3in); height 4.17m (13ft 8in)
Weight: 5488kg (12,100lb) loaded
Armament: 6 x 12.7mm (0.5in) MGs in wings, plus up to 2 x 454kg (1000lb) bombs

IX Tactical Air Command

June 1944 – May 1945

The IX Tactical Air Command was the subordinate formation within the IX Fighter Command, tasked with the support of the US First Army. In this task, it operated from quickly prepared advanced landing grounds close behind the front for minimum response times.

ESTABLISHED LATE in 1943, the IX Tactical Air Command was led by Major-General Elwood Quesada, and comprised three fighter-bomber wings and one reconnaissance group.

In June 1944, on the eve of the Allied invasion of France, the IX TAC had three fighter-bomber wings. The 70th Fighter Wing comprised Colonel George Wertenbaker's Ibsley-based 48th Fighter Group (492nd, 493rd and 494th Fighter Squadrons) flying the P-47 Thunderbolt, Colonel Charles Young's Stoney Cross-based 367th Fighter Group (392nd, 393rd and 394th Fighter Squadrons) flying the P-38 Lightning, Colonel Bingham Kleine's Bisterne-based 371st Fighter Group (404th, 405th and 406th Fighter Squadrons) flying the P-47 Thunderbolt, and Colonel Clinton Wasem's Warmwell-based 474th Fighter Group (428th, 429th and 430th Fighter Squadrons) flying the P-38 Lightning.

The 71st Fighter Wing comprised Lieutenant-Colonel Norman Holt's Thruxton-based 366th Fighter Group (389th, 390th and 391st Fighter Squadrons) flying the P-47 Thunderbolt, Colonel Gil Meyers' Chilboton-based 368th Fighter Group (395th, 396th and 397th Fighter Squadrons) flying the P-47 Thunderbolt, and Colonel Seth McKee's Andover-based 370th Fighter Group (401st, 402nd and 385th Fighter Squadrons) flying the P-38 Lightning.

The 84th Fighter Wing comprised Colonel William Greenfield's Lymington-based 50th Fighter Group (10th, 81st and 313rd Fighter Squadrons) flying the P-47 Thunderbolt, Colonel Ray Steckers's Beaulieu-based 365th Fighter Group (386th, 387th and 388th Fighter Squadrons) flying the P-47

ORGANIZATION

IX Tactical Air Command
June 1944

- 70th Fighter Wing
- 71st Fighter Wing
- 84th Fighter Wing
- 67th Tactical Reconnaissance Group

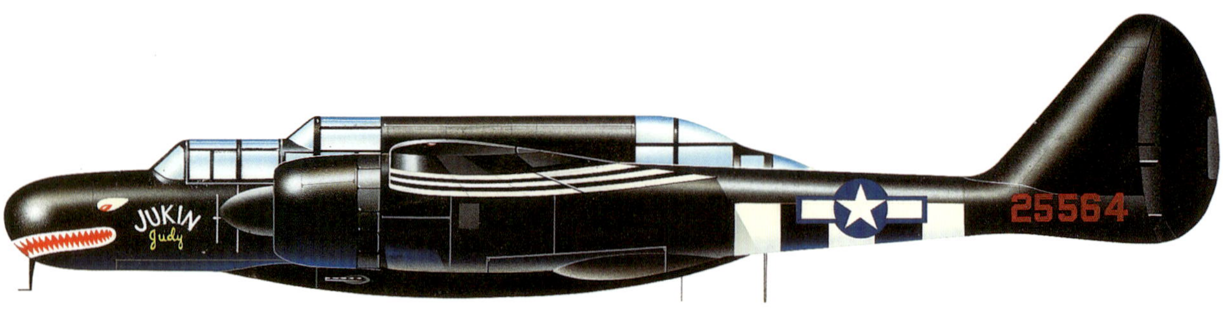

Specifications

Crew: 2/3
Powerplant: 2 x 1678kW (2250hp) Pratt & Whitney R-2800-65 18 cylinder radial engines
Maximum speed: 594km/h (369mph)
Range: 3058km (1900 miles)
Service ceiling: 10.090m (33,100ft)
Dimensions: span 20.12m (66ft); length 14.91m (48ft 11in); height 4.46m (14ft 8in)
Weight: 15,513kg (34,200lb) maximum takeoff
Armament: 4 x 20mm (0.8in) cannon in underside of forward fuselage

▲ **Northrop P-61A Black Widow**

422nd Night-Fighter Squadron / IX Fighter Command, Scorton, 1944

One of two night-fighter squadrons, the other being the 425th Night-Fighter Squadron, which reported directly to the headquarters of the IX Fighter Command, the 422nd Night-Fighter Squadron worked up at Scorton in Yorkshire before moving south to its operational base at Charmy Down.

367TH & 474TH FIGHTER GROUPS, 70TH FIGHTER WING			
Unit	**Type**	**Strength**	**Serviceable**
392nd FS	P-38	16	n/a
393rd FS	P-38	16	n/a
394th FS	P-38	16	n/a
428th FS	P-38	16	n/a
429th FS	P-38	16	n/a
430th FS	P-38	16	n/a

48TH & 371ST FIGHTER GROUPS, 70TH FIGHTER WING			
Unit	**Type**	**Strength**	**Serviceable**
492nd FS	P-47	16	n/a
493rd FS	P-47	16	n/a
494th FS	P-47	16	n/a
404th FS	P-47	16	n/a
405th FS	P-47	16	n/a
406th FS	P-47	16	n/a

Thunderbolt, Colonel Carol McColpin's Winkton-based 404th Fighter Group (506th, 507th and 508th Fighter Squadrons) flying the P-47 Thunderbolt, and Colonel Robert Delashew's Christchurch-based 405th Fighter Group (509th, 510th and 511th Fighter Squadrons) flying the P-47 Thunderbolt.

Reconnaissance elements

The IX TAC was completed by its own organic tactical and photo-reconnaissance element in the form of Colonel George Peck's 67th Tactical Reconnaissance Group, which was based largely at Chalgrove and Middle Wallop, and comprised the 107th and 109th Tactical Reconnaissance Squadrons flying the P-51 Mustang, and the 30th and 33rd Photo-Reconnaissance Squadrons flying the F-5 reconnaissance version of the P-38 Lightning.

After providing support for the initial stages of the Allied invasion from across the Channel, these wings moved to France as and when there were facilities available, and moved forward according to the general Allied advance.

Ground attack

Here, a Ninth Air Force Lightning strafes a train in German-occupied Europe in late 1944. Note the' invasion stripes' on the wings.

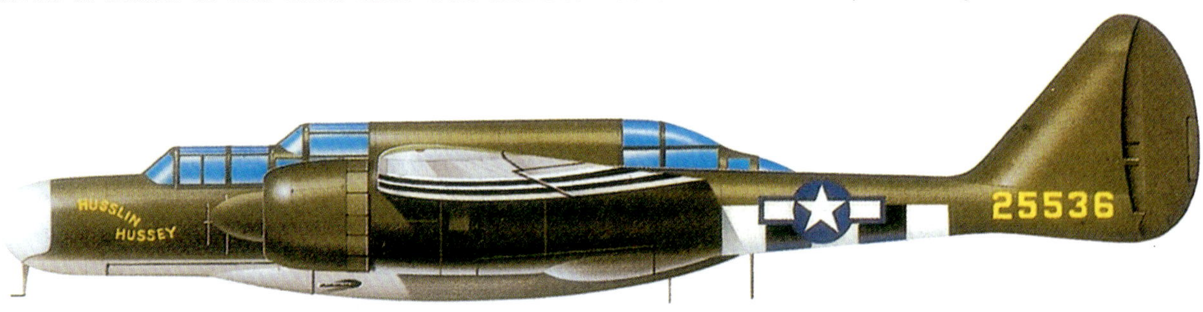

Specifications

Crew: 2/3
Powerplant: 2 x 1678kW (2250hp) Pratt & Whitney R-2800-65 18 cylinder radial engines
Maximum speed: 594km/h (369mph)
Range: 3058km (1900 miles)
Service ceiling: 10.090m (33,100ft)
Dimensions: span 20.12m (66ft); length 14.91m (48ft 11in); height 4.46m (14ft 8in)
Weight: 15,513kg (34,200lb) maximum takeoff
Armament: 4 x 20mm (0.8in) cannon in underside of forward fuselage

Northrop P-61A Black Widow

422nd Night-Fighter Squadron / IX Fighter Command, Scorton, summer 1944

The 422nd Night-Fighter Squadron was the first unit equipped with the P-61 to reach the UK, and flew its first operational sortie in July 1944 from Hurn, near Bournemouth.

Specifications

Crew: 1
Powerplant: 1044kW (1400hp) Packard V-1650-3 Rolls-Royce Merlin engine
Maximum speed: 690km/h (430mph)
Range: 3540km (2200 miles)
Service ceiling: 12,649m (41,500ft)
Dimensions: span 11.27m (37ft); length 9.84m (32ft 4in); height 4.15m (13ft 8in)
Weight: 4173kg (9200lb) loaded
Armament: 6 x 12.7mm (0.5in) MGs

▲ **North American P-51B Mustang**

356th Fighter Squadron / 354th Fighter Group, UK, 1944

This was the second aeroplane with this name flown by Major James Howard, the commander of the 356th Fighter Squadron. This officer had previously flown with the American Volunteer Group in China, hence the six 'meatball' markings above the six swastika kill markings below the cockpit.

XIX Tactical Air Command

June 1944 – May 1945

The XIX Tactical Air Command was the formation entrusted with the provision of tactical air support for the US Third Army. This was the fastest-moving of the US armies, and the XIX TAC therefore became very adept in the art of fast relocation farther forward.

Established in 1944, the XIX Tactical Air Command was led by Major-General Otto Weyland, and initially comprised two fighter-bomber wings and one reconnaissance group.

Thus as General George S. Patton's newly activated Third Army began to move south-east through France in an advance that would eventually take its spearhads into western Czechoslovakia by May 1945, the XIX TAC had two fighter-bomber wings.

The 100th Fighter Wing comprised Colonel George Bickell's 48th Fighter Group (353rd, 355th and 366th Fighter Squadrons) originally based at Lashenden and flying the P-51 Mustang, Colonel Cecil Wells' 358th Fighter Group (365th, 366th and 367th Fighter Squadrons) originally based at High Halden and flying the P-47 Thunderbolt, Colonel Morton Magaffin's 362nd Fighter Group (377th, 378th and 379th Fighter Squadrons) originally based at Headcorn and flying the P-47 Thunderbolt, and Colonel Jim Tipton's 363rd Fighter Group (380th, 381st and 382nd Fighter Squadrons) originally based at Staplehurst and flying the P-51 Mustang.

ORGANIZATION

XIX Tactical Air Command
September 1944

100th Fighter Wing | 303rd Fighter Wing | 10th Photo Group

303rd Fighter Wing

The 303rd Fighter Wing comprised Colonel Lewis Curry's 36th Fighter Group (22nd, 23rd and 53rd Fighter Squadrons), originally based at Kingsnorth and flying the P-47 Thunderbolt, Colonel William Schwarz's 373rd Fighter Group (410th, 411th and 412th Fighter Squadrons), first based at Woodchurch and flying the P-47 Thunderbolt, and Colonel Anthony Grossetta's 40th Fighter Group (412th, 513th and 514th Fighter Squadrons) originally based at Ashford and flying the P-47 Thunderbolt. On 15 September, the 303rd Fighter Wing was transferred to the XXIX TAC.

The XIX TAC was completed by its own organic tactical and photo reconnaissance element in the form of Colonel William Reid's 10th Photo Group, which comprised the 12th and 15th Tactical Reconnaissance Squadrons flying the P-51 Mustang, and the 31st and 34th Photo-Reconnaissance Squadrons flying the P-38 Lightning. All these units soon moved from southern England to France and then points farther east and south.

NINTH AAF LOSSES IN EUROPE (NOVEMBER 1943 – MAY 1945)

Aircraft type	Number
Douglas A-20/F-3 Havoc	57
Douglas A-26 Invader	35
Martin B-26 Marauder	179
Douglas C-47	92
Lockheed P-38/F-5 Lightning	213
Republic P-47 Thunderbolt	616
North American P-51/F-6 Mustang	233
Northrop P-61 Black Widow	4

▲ North American P-51B Mustang

353rd Fighter Squadron / 354th Fighter Group, Lashenden, summer 1944

This was the second aeroplane flown by Captain Donald M. 'Buzz' Beerbower to carry the name 'Bonnie B'.

Specifications

Crew: 1
Powerplant: 1044kW (1400hp) Packard V-1650-3 Rolls-Royce Merlin engine
Maximum speed: 690km/h (430mph)
Range: 3540km (2200 miles)
Service ceiling: 12,649m (41,500ft)
Dimensions: span 11.27m (37ft); length 9.84m (32ft 4in); height 4.15m (13ft 8in)
Weight: 4173kg (9200lb) loaded
Armament: 6 x 12.7mm (0.5in) MGs

Specifications

Crew: 1
Powerplant: 1264kW (1695hp) Packard Merlin V-1650-7 V-12 piston engine
Maximum speed: 703km/h (437mph)
Range: 3347km (2080 miles)
Service ceiling: 12,770m (41,900ft)
Dimensions: span 11.28m (37ft); length 9.83m (32ft 3in); height 4.17m (13ft 8in)
Weight: 5488kg (12,100lb) loaded
Armament: 6 x 12.7mm (0.5in) MGs in wings

▲ North American F-6D Mustang

15th Tactical Reconnaissance Squadron / 10th Photographic Group / XIX Tactical Air Command, northern France, late summer 1944

This was the reconnaissance development of the P-51D, with cameras in the rear fuselage to the rear of the radiator installation.

Specifications

Crew: 1
Powerplant: 1044kW (1400hp) Packard V-1650-3 Rolls-Royce Merlin engine
Maximum speed: 690km/h (430mph)
Range: 3540km (2200 miles)
Service ceiling: 12,649m (41,500ft)
Dimensions: span 11.27m (37ft); length 9.84m (32ft 4in); height 4.15m (13ft 8in)
Weight: 4173kg (9200lb) loaded
Armament: 6 x 12.7mm (0.5in) MGs

▲ North American P-51B Mustang

382nd Fighter Squadron / 363rd Fighter Group / XIX Tactical Air Command, northern France, summer 1944

This was the fighter flown by Robert McGee after the move of the 363rd Fighter Group to France, as indicated by the yellow spinner adopted at this time.

Specifications

Crew: 1
Powerplant: 1264kW (1695hp) Packard Merlin V-1650-7 V-12 piston engine
Maximum speed: 703km/h (437mph)
Range: 3347km (2080 miles)
Service ceiling: 12,770m (41,900ft)
Dimensions: span 11.28m (37ft); length 9.83m (32ft 3in); height 4.17m (13ft 8in)
Weight: 5488kg (12,100lb) loaded
Armament: 6 x 12.7mm (0.5in) MGs in wings, plus up to 2 x 454kg (1000lb) bombs

▲ North American P-51D Mustang

402nd Fighter Squadron / 370th Fighter Group / IX Tactical Air Command, France, late 1944

This was the aeroplane used almost exclusively in the fighter-bomber role by the pilot Robert Bohna.

Specifications

Crew: 1
Powerplant: 1044kW (1400hp) Packard V-1650-3 Rolls-Royce Merlin engine
Maximum speed: 690km/h (430mph)
Range: 3540km (2200 miles)
Service ceiling: 12,649m (41,500ft)
Dimensions: span 11.27m (37ft); length 9.84m (32ft 4in); height 4.15m (13ft 8in)
Weight: 4173kg (9200lb) loaded
Armament: 6 x 12.7mm (0.5in) MGs

▲ North American F-6C Mustang

15th Tactical Reconnaissance Squadron / 10th Photographic Group, France, autumn 1944

The aeroplane of John Hoefker, who ended World War II with 10.5 victories. He was shot down on the German side of the frontline in December 1944 but escaped capture to return to his squadron.

Specifications

Crew: 1
Powerplant: 1264kW (1695hp) Packard Merlin V-1650-7 V-12 piston engine
Maximum speed: 703km/h (437mph)
Range: 3347km (2080 miles)
Service ceiling: 12,770m (41,900ft)
Dimensions: span 11.28m (37ft); length 9.83m (32ft 3in); height 4.17m (13ft 8in)
Weight: 5488kg (12,100lb) loaded
Armament: 6 x 12.7mm (0.5in) MGs in wings, plus up to 2 x 454kg (1000lb) bombs

▲ **North American P-51D Mustang**

356th Fighter Squadron / 354th Fighter Group / XIX Tactical Air Command, 1944–45

This was the aeroplane of Lieutenant-Colonel Richard Turner, commander of the 356th Fighter Squadron.

XXIX Tactical Air Command

September 1944 – May 1945

Activated only in September 1944 as the Allied armies swept east through northern France into Belgium and toward western Germany, the XXIX Tactical Air Command was allocated the task of providing the US Ninth Amy with tactical air support.

It was during September 1944 that the XXIX TAC was officially activated under the command of Brigadier-General Richard E. Nugent for the tactical air support of the US Ninth Army. At this stage, the command's assets were only the 303rd Fighter Wing transferred from the XIX TAC, and the 363rd Reconnaissance Group.

ORGANIZATION

XXIX Tactical Air Command
September 1944

303rd Fighter Wing | 363rd Reconnaissance Group

303rd Fighter Wing

Operating from a number of bases in parts of France liberated from German occupation, the 303rd Fighter Wing comprised Colonel Lewis Curry's Athis-based 36th Fighter Group (22nd, 23rd and 53rd Fighter Squadrons) flying the P-47 Thunderbolt, Colonel William Schwarz's Reims-based 373rd Fighter Group (410th, 411th and 412th Fighter Squadrons) flying the P-47 Thunderbolt, and Colonel Anthony Grossetta's Mourmelon le Grand-based 40th Fighter Group (512th, 513th and 514th Fighter Squadrons) also flying the P-47 Thunderbolt.

In the following month, having received units from the IX and XIX TACs, the XXIX TAC became operational in support of the US Ninth Army, and one of the new formation's most important assets was Major Richard Leghorn's 33rd Photo Reconnaissance Squadron, equipped with the F-5 photo-reconnaissance derivative of the P-38 Lightning. which had been received from the IX TAC's 67th Reconnaissance Group and remained with the 363rd Tactical Reconnaissance Group for the rest of World War II.

The capabilities of the XXIX TAC, which remained smaller than the other two tactical air commands of the IX Fighter Command, were revealed on the first day of *Grenade*, the Ninth Army's crossing of the Roer river on 23 February 1945. For this undertaking, the XXIX TAC could call on its reconnaissance group and, more immediately significant, the armed power of its five groups of fighter-bombers, totalling some 375 aircraft. The Americans crossed the river without undue difficulty, but had then to face German counterattacks.

The Ninth Army's artillery was very effective, and was ably supported in this task by the Thunderbolt squadrons of the XXIX TAC, which operated at very low level to drop their bombs with commendable accuracy before strafing the hapless Germans soldiers with their devastating batteries of eight 12.7mm (0.5in) Browning machine guns.

'Plunder'

Much the same level of capability was revealed in the days before the launch of the *Plunder* offensive to take the Allied 21st Army Group, including the US Ninth Army, across the great Rhine river into Germany proper on 23 March 1945. As part of the interdiction effort designed to isolate the German

ORGANIZATION

303rd Fighter Wing
Sept 1944 – May 1945

36th Fighter Group	373rd Fighter Group	40th Fighter Group
22nd, 23rd and 53rd Fighter Squadrons	410th, 411th and 412th Fighter Squadrons	512th, 513th and 514th Fighter Squadrons

Ground attack master

Fitted with eight 12.7mm (0.5in) machine guns and capable of carrying bombs and rockets, the P-47 was the ideal ground attack platform.

Republic P-47D Thunderbolt

512th Fighter Squadron / 406th Fighter Group, Nordholz, Germany, summer 1945

The red/yellow/red band round the rear fuselage indicates that this unit had been allocated to the air element of the occupation forces.

Specifications

Crew: 1
Powerplant: 1891kW (2535hp) Pratt & Whitney R-2800-59W Double Wasp
Maximum speed: 697km/h (433mph)
Range: 3060km (1900 miles) with drop tanks
Service ceiling: 12,495m (41,000ft)
Dimensions: span 12.42m (40ft 9in); length 11.02m (36ft 2in); height 4.47m (14ft 8in)
Weight: 7938kg (17,500lb) maximum takeoff
Armament: 8 x 12.7mm (0.5in) MGs in wings, plus provision for 1134kg (2500lb) external bombs or rockets

▲ **P-51B Mustang**

To many pilots, the P-51B was the finest variant of the P-51 Mustang. When fitted with the British 'Malcolm hood', it was lighter and faster, and handled more crisply, than the bubble-canopied P-51D. This P-51B flew with the 354th Fighter Squadron.

side of the river, the Allied air forces had since the middle of February been running a major bombing effort. This was designed to seal the Ruhr from the rest of Germany by destroying rail bridges and viaducts, and by attacking canal traffic along a broad arc running from Bremen near the North Sea south and south-west around the eastern edge of the Ruhr to the Rhine south of the industrial region.

Attacks were aimed at critical German communications centres, railway marshalling yards, supply dumps, industrial plants and any other comparable targets. From mid-February to 21 March, much of the Allied air strength had been focused on this task whenever other operations permitted. Heavy and medium bombers made some 1792 sorties against 17 rail bridges and viaducts along the arc round the Ruhr. By 21 March, 10 of the bridges had been destroyed and five others rendered unusable. The destruction of these bridges significantly impaired German movement.

Final push

The fighters and fighter-bombers of the British 2nd TAF and XXIX TAC now joined the Ruhr campaign. Most of the 7311 sorties flown by these pilots between 11 and 21 March were directed against the rail and road systems of the Ruhr. In the last three days before the operation's start, some 2000 medium bombers of the US 9th Bombardment Division struck at communications centres, rail yards and flak positions, and thus the area was effectively isolated.

US Twelfth Air Force

1942–45

The US Twelfth Air Force came into being as a direct result of the decision by US planners that proposed operations in north-west Africa required more air support than was currently available. The formation then went on to vital service in the Italian campaign.

THE ORIGINS OF THE US Twelfth Air Force date to the meetings held between British and US planning staffs in mid-1942. The aim was to develop the strategy that would see the introduction of US forces into combat against the Germans for the first time, namely the *Torch* amphibious landings in French north-west Africa and the subsequent eastward advance to Tunisia. This was an operation and subsequent campaign of such size and complexity that it would require considerable air support. Indeed, recognizing that a new command organization would have to be established to control the personnel and equipment involved, the planners decided to create a new US Twelfth Air Force.

The new Twelfth Air Force was activated on 20 August 1942 at Bolling Field, Maryland, and on 23 September Brigadier-General 'Jimmy' Doolittle assumed command with Colonel Hoyt S. Vandenberg as chief of staff. Time for the preparation of the new air force was short. In fact, it was only four

ORGANIZATION

US Twelfth Air Force
mid-1944

- XII Bomber Command
- XII Air Support Command
- XXII Fighter Command

USAAF GROUP STANDARD ORGANIZATION AND STRENGTH

Unit	No. of aircraft	No. of crews
Medium bomber	96	96
Light bomber	96	96
Single-engine fighter	110–125	108–125
Twin-engine fighter	110–125	108–125
Night-fighter	18	16
Tactical recce	27	23
Photo recce	24	21

Bell P-39L Airacobra

93rd Fighter Squadron / 81st Fighter Group

The wheel turned full circle as the US forces flew Airacobra aircraft initially delivered to the UK as Airacobra Mk I warplanes. The Airacobra's strength and heavy armament made it a good attack fighter.

Specifications

Crew: 1
Powerplant: 895kW (1200hp) Allison V-1710-85 liquid-cooled V-12
Maximum speed: 605km/h (376mph)
Range: 483km (300 miles)
Service ceiling: 11,665m (38,270ft)
Dimensions: span 10.36m (34ft 0in); length 9.195m (30ft 2in); height 3.785m (12ft 5in)
Weight: 3992kg (8800lb) loaded
Armament: 1 x 37mm (1.46in) M4 cannon, 2 x 12.7mm (0.5in) Browning MGs, 4 x 7.62mm (0.3in) Browning MGs, and 227kg (500lb) of bombs

months before the Twelfth Air Force was committed to active service with the descent of Allied forces on north-west Africa on 8 November 1942. The Twelfth Air Force was very active in the period leading to the final defeat of the Axis forces in North Africa during May 1943, and after this it was reduced for further service in the Mediterranean theatre. The US military leadership was basically opposed to the continuance of operations in this theatre (believing it to be a diversion from the primary aim of defeating the Germans by the most direct assault from northern France into Germany), but was persuaded to agree to operations that saw the Twelfth Air Force in action over Sicily, Italy and southern France.

By May 1945, the Twelfth had flown some 430,681 sorties, dropped 220,630 tonnes (217,156 tons) of bombs, claimed destruction of 2857 aircraft and lost 2667 of its own. The Twelfth Air Force was eventually deactivated at Florence, Italy, on 31 August 1945.

▲ **Supermarine Spitfire Mk VC**

308th Fighter Squadron / 31st Fighter Group

Fitted under the nose with a Vokes filter to prevent North African dust and grit being drawn into the engine, this Spitfire Mk VC was one of a modest number of British aircraft operated in North Africa by the US air forces.

Specifications

Crew: 1

Powerplant: 1096kW (1470hp) Rolls-Royce Merlin 50 engine

Maximum speed: 594km/h (369mph)

Range: 1827km (1135 miles)

Service ceiling: 11,125m (36,500ft)

Dimensions: span 11.23m (36ft 10in); length 9.12m (29ft 11in); height 3.02m (9ft 11in)

Weight: 2911kg (6417lb) loaded

Armament: 4 x 7.7mm (0.303in) MGs and 2 x 20mm (0.8in) cannons in wings

North Africa and the Mediterranean

1942–45

The Mediterranean theatre was never anything more than a sideshow to the US armed forces. As a result of British pressure, however, the Americans agreed to an extension of the success in North Africa to begin operations against Italy, first on Sicily and then the mainland.

After the defeat of the Axis forces in North Africa, the Allies decided, despite US reluctance, to pursue them into Italy. The first step was the *Husky* invasion of Sicily by the US Seventh and British Eighth Armies in July 1943. The Allies then prepared to assault the mainland, starting with the *Avalanche* descent on Salerno in September 1943 by the US Fifth and British Eighth Armies. Italy secured an armistice just as this operation was launched, but the Germans decided to fight.

Very ably led and including a number of high-grade formations, the German armies were able to make the best of Italy's formidable campaigning terrain to fight a series of hard-fought delaying actions as they were slowly and very bloodily driven north along the Italian peninsula.

Defensive barrier

The British were largely on the east and the Americans on the west. The Apennine mountains made the basis of superb defensive lines, which the Germans held strongly before falling back in good order, and the Allies were never able to outflank these defences in any significant manner.

Nevertheless, the Fifth Army captured Rome in June 1944, and with potent and overwheming air support, fought north to eventual victory in April 1945.

▲ Grumman F4F-4 Wildcat

VGR-28 / USS* Suwannee, *northwest Africa, November 1942

Used in the reconnaissance fighter role, this US Navy warplane reflects the removal of red from the US national insignia and the adoption of a large yellow surround to lessen the chances of confusion with Axis markings.

Specifications

Crew: 1
Powerplant: 895kW (1200hp) Pratt & Whitney R-1830-86 14-cylinder engine
Maximum speed: 507km/h (315mph)
Range: 1851km (1150 miles)
Service ceiling: 12,010m (39,400ft)
Dimensions: span 11.58m 38ft); length 8.76m (28ft 8in); height 2.81m (9ft 3in)
Weight: 3607kg (7952lb) loaded
Armament: 6 x 12.5mm (0.5in) MGs in wings; 2 x 45kg (100lb) bombs

Specifications

Crew: 1
Powerplant: 895kW (1200hp) Allison piston engine
Maximum speed: 563km/h (350mph)
Range: 1738km (1080 miles)
Service ceiling: 9450m (31,000ft)
Dimensions: span 11.36m (37ft 4in); length 10.16m (33ft 4in); height 3.76m (12ft 4in)
Weight: 3511kg (7740lb) loaded
Armament: 4 x 12.7mm (0.5in) Browning MGs in wings

▲ Curtiss P-40L Warhawk

325th Fighter Group, Tunisia, 1943

This was the aeroplane of the group's commander, Lieutenant-Colonel Gordon H. Austin. The aeroplane was completed in the standard 'sand and spinach' camouflage of the theatre below national, unit and personal markings.

Bomber escort – Italian Campaign
1942–45

The US Twelfth Air Force was brought into being as a balanced air formation able to satisfy all the requirements of the US Army's senior commanders in North Africa. The formation therefore included all of the elements for modern air warfare over and beyond the land battlefield.

The Twelfth Air Force was planned from the outset as a balanced air formation able to provide all levels of bombing, air support and fighter capabilities complemented by a small but adequate reconnaissance capability. The main weight of its offensive capability was, of course, its bomber arm, and the XII Bomber Command was constituted on 26 February 1942 and activated on 13 March at the MacDill base in Florida. It was assigned to the Twelfth Air Force in August and transferred, without personnel and equipment, to High Wycombe in England, where the command was re-established before being moved to Tafaraoui in Algeria on 22 November 1942 to support the US and British ground forces after the *Torch* landings.

The XII Bomber Command was operational in the Mediterranean theatre until 1 November 1943, when most of its personnel were withdrawn. The command received more men in January 1944 and served in combat until 1 March 1944, but was disbanded in Corsica on 10 June 1944.

ORGANIZATION

XII Tactical Air Command
June 1943

- 63rd Fighter Wing
- 64th Fighter Wing
- 87th Fighter Wing

Bomber escort

Known elements of the XII Bomber Command were the 5th Bombardment Wing comprising five bombardment groups with the B-17, two bombardment groups with the B-24, one bombardment group with the A-20 and A-26, two

US Fighter Squadron, 1942–43

The standard organization of the USAAF fighter and fighter-bomber squadron in the Mediterranean campaign was based on a strength of 16 aircraft tactically disposed as four flights each with four aircraft, the aircraft of each flight operating in two pairs in leader and wingman combinations. The arrangement worked well, and remained essentially unaltered for the rest of World War II.

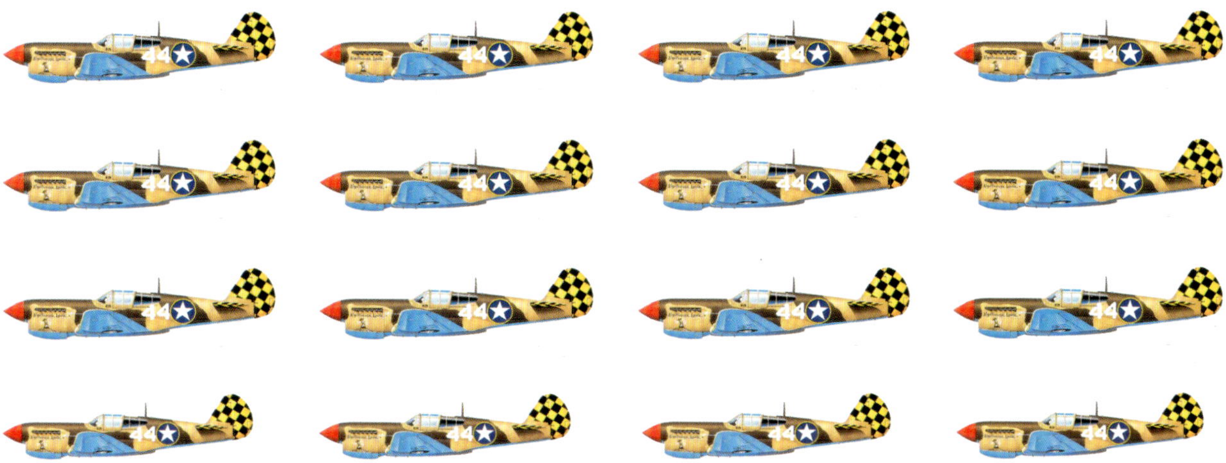

fighter groups with the P-38, one fighter group with the P-40, and one reconnaissance group with the P-38, P-39, P-40, P-51, A-20, A-36, B-17 and B-24. The 42nd Bombardment Wing comprised three bombardment groups with the B-26, one fighter group with the P-38 and one fighter group with the P-40; the 47th Bombardment Wing comprised two bombardment groups with the B-24, two bombardment groups with the B-25 and three bombardment groups with the B-26, one fighter group with the P-38, and three fighter groups with the P-40. The 57th Bombardment Wing comprised five bombardment groups with the B-25, one bombardment with the A-20 and A-26, and two fighter groups with the P-40.

USAAF COMBAT GROUPS IN THE MEDITERRANEAN THEATRE			
Date	**Fighter**	**Light/medium bomber**	**Heavy bomber**
December 1942	10	6	3
December 1943	13	8	9
December 1944	12	8	19
March 1945	10	8	19

North American P-51A Mustang

522nd Fighter-Bomber Squadron

Optimized for the long-range role, the P-51A was a fighter with good speed but was limited in the air-combat role by the poor performance of its Allison V-1710 liquid-cooled V-12 engine. As Merlin-engined Mustangs with better altitude capability were introduced, the P-51A became a useful fighter-bomber.

Specifications

Crew: 1
Powerplant: 895kW (1200hp) Allison V-1710-81 engine
Maximum speed: 690km/h (430mph)
Range: 3540km (2200 miles)
Service ceiling: 12,649m (41,500ft)
Dimensions: span 11.27m (37ft); length 9.84m (32ft 4in); height 4.15m (13ft 8in)
Weight: 4173kg (9200lb) loaded
Armament: 4 x 12.7mm (0.5in) MGs

Apaches in the Desert

The A36A Apache was first used by the USAAF in the North African campaign in early 1943. This aircraft was rejected for widespread production in favour of the Mustang, but it was instrumental in the development of later Mustang variants.

Specifications

Crew: 1

Powerplant: 988kW (1325hp) Allison V-1710-87 liquid-cooled piston V12 engine

Maximum speed: 590km/h (365mph)

Range: 885km (550 miles)

Service ceiling: 7650m (25,100ft)

Dimensions: span 11.28m (37ft 1in); length 9.83m (32ft 3in); height 3.71m (12ft 2in)

Weight: 4535kg (10001b) loaded

Armament: 6 x 12.7mm (0.50in) M2 Browning MGs in wings

▲ **North American A-36A Apache**

524th Fighter-Bomber Squadron, Italy, 1943

The A-36A was the attack fighter brother of the P-51 Mustang fighter, as suggested by the indifferent altitude performance of its Allison V-1710 liquid-cooled V-12 engine. The type was flown in Italy by the 27th and 86th Fighter-Bomber Groups. Only 500 such aircraft were built.

XII Tactical Air Command

1942–45

The formation that finally became the XII Tactical Air Command was initially created as the XII Ground Air Support Command. It was intended to provide the US forces of the Allied First Army in North Africa with a powerful yet flexible air support.

THE XII TACTICAL Air Command (TAC) was constituted on 10 September 1942 with the initial designation XII Ground Air Support Command, and was activated on 17 September. It was allocated to the new Twelfth Air Force, receiving the revised designation XII Air Support Command before finally becoming the XII TAC in the course of April 1944.

The XII Air Support Command arrived in French Morocco from 9 November 1942 as the major part of the US air component that supported the Allied *Torch* landings.

Mediterranean base

The XII TAC was operational in the Mediterranean and European theatres until May 1945, and then remained in Europe as part of the occupation force of the United States Air Forces in Europe (USAFE). It was deactivated at Bad Kissingen in the US occupation zone of Germany on 10 November 1947.

ORGANIZATION

XII Tactical Air Command
June 1945

5th Bombardment Wing	42nd Bombardment Wing	57th Bombardment Wing	64th Fighter Wing

Known XII TAC components included three bombardment wings transferred from the XII Bomber Command when this was deactivated in the course of June 1944. These three units were the 5th Bombardment Wing for the heavy bomber role with the 2nd Bombardment Group (1944–45), the 97th Bombardment Group (1944–45), the 99th Bombardment Group (1944–45), the 301st Bombardment Group (1944–45), the 463rd Bombardment Group (1944–45) and the 483rd

Bombardment Group (1944–45), all flying the B-17 Flying Fortress four-engine bomber. For the medium bomber role, the 42nd Bombardment Wing with the 17th Bombardment Group (1944–45) and 320th Bombardment Group (1944–45) both flying the B-26 Marauder twin-engine bomber; and the 57th Bombardment Wing also for the medium bomber role with the 310th Bombardment Group (1944–45), 319th Bombardment Group (1944–45), 321st Bombardment Group (1944–45) and 340th Bombardment Group (1944–45), all flying the B-25 Mitchell twin-engine bomber.

▲ Supermarine Spitfire Mk VIII

308th Fighter Squadron / 31st Fighter Group, Italy, 1944

This was the aeroplane of 1st Lieutenant L. P. Molland, commanding officer of the 308th Fighter Squadron.

Specifications

Crew: 1
Powerplant: 1170kW (1565hp) 12-cylinder Rolls-Royce Merlin 61 engine
Maximum speed: 642km/h (410mph)
Range: 698km (435 miles) on internal fuel tanks
Service ceiling: 12,650m (41,500ft)
Dimensions: span 11.23m (36ft 10in); length 9.47m (31ft 1in); height 3.86m (12ft 8in)
Weight: 3343kg (7370lb) loaded
Armament: 4 x 7.7mm (0.303in) MGs and 2 x 20mm (0.8in) cannons

Specifications

Crew: 2
Powerplant: 2 x 1230kW (1650hp) Bristol Hercules VI radial piston engines
Maximum speed: 488km/h (303mph)
Range: 2366km (1470 miles)
Service ceiling: 4570m (15,000ft)
Dimensions: span 17.63m (57ft 10in); length 12.70m (41ft 8in); height 4.83m (15ft 10in)
Weight: 11,431kg (25,200lb) loaded
Armament: 4 x 20mm (0.8in) cannon and 6 x 7.7mm (.303in) MGs in wings; 1 x 7.7mm (.303in) Vickers MG in dorsal turret

▲ Bristol Beaufighter Mk VIF

416th Night-Fighter Squadron, Corsica, 1943–44

Another British type that saw limited service with the USAAF was the Beaufighter, seen here in its Mk VIF form as a radar-equipped night-fighter.

Post-war service

The last of the XII TAC's components was the 64th Fighter Wing, which was added for service on post-war USAFE occupation duties. This wing comprised the 27th Fighter Group (1946–47), 36th Fighter Group (1945–46), 86th Fighter Group (1945–46), 324th Fighter Group (1945) and 406th Fighter Group (1945–46) all flying the P-47 Thunderbolt single-engine heavy fighter; and the 52nd Fighter Group (1946–47), 354th Fighter Group (1945–46) and 355th Fighter Group (1945), all flying the P-51 Mustang single-engined multi-role fighter.

Specifications

Crew: 1
Powerplant: 1891kW (2535hp) Pratt & Whitney R-2800-59W Double Wasp
Maximum speed: 697km/h (433mph)
Range: 3060km (1900 miles) with drop tanks
Service ceiling: 12,495m (41,000ft)
Dimensions: span 12.42m (40ft 9in); length 11.02m (36ft 2in); height 4.47m (14ft 8in)
Weight: 7938kg (17,500lb) maximum takeoff
Armament: 8 x 12.7mm (0.5in) MGs in wings, plus provision for 1134kg (2500lb) external bombs or rockets

▲ Republic P-47D Thunderbolt

86th Fighter Squadron / 79th Fighter Group, Fano, Italy, February 1945

While serving in Italy, the 79th Fighter Group used a non-standard group designation system (x plus a number), which it had first adopted in the USA before transfer overseas.

Specifications

Crew: 1
Powerplant: 1044kW (1400hp) Packard V-1650-3 Rolls-Royce Merlin engine
Maximum speed: 690km/h (430mph)
Range: 3540km (2200 miles)
Service ceiling: 12,649m (41,500ft)
Dimensions: span 11.27m (37ft); length 9.84m (32ft 4in); height 4.15m (13ft 8in)
Weight: 4173kg (9200lb) loaded
Armament: 6 x 12.7mm (0.5in) MGs

▲ North American P-51C Mustang

5th Fighter Squadron / 52nd Fighter Group, Italy, 1944

This was the aeroplane of Lieutenant Calvin D. Allen, Jr.

▲ **Republic P-47D Thunderbolt**

USAAF, European Theatre, 1944

This is the original pattern of P-47D with a framed canopy sliding to the rear over the 'razorback' turtledeck upper decking of the rear fuselage. Even with drop tanks, the P-47D possessed range that was adequate rather than exceptional.

Specifications

Crew: 1
Powerplant: 1891kW (2535hp) Pratt & Whitney R-2800-59W Double Wasp
Maximum speed: 697km/h (433mph)
Range: 3060km (1900 miles) with drop tanks
Service ceiling: 12,495m (41,000ft)
Dimensions: span 12.42m (40ft 9in); length 11.02m (36ft 2in); height 4.47m (14ft 8in)
Weight: 7938kg (17,500lb) maximum takeoff
Armament: 8 x 12.7mm (0.5in) MGs in wings, plus provision for 1134kg (2500lb) external bombs or rockets

XXII Tactical Air Command

1942–45

Despite its several names, the XXII TAC provided the US Twelfth Air Force with very useful ground-attack and close support capabilities, and also the wherewithal to provide fighter escort for its bomber forces.

THE LAST MAJOR component of the US Twelfth Air Force, and optimized for the tactical role, was the XXII TAC. This was constituted on the same day as the XII Bomber Command, on 26 February 1942, and was activated on 5 March, thus predating the XII Ground Air Support Command, which later became the XII TAC. The XXII TAC was redesignated as the XII Fighter Command in May 1942, but reverted to its original designation as the XXII TAC during the course of November 1944.

ORGANIZATION

XXII Tactical Air Command
June 1945

63rd Fighter Wing | **87th Fighter Wing** | **64th Fighter Wing**

North African posting

The XII Fighter Command was allocated to the Twelfth Air Force in August 1942. Establishedt at Wattisham in England during September 1942, it moved to its designated operational area in North Africa, reaching Tafaraoui in French Algeria on 8 November 1942 as part of the air component for the Allied forces involved in the *Torch* landings in French north-west Africa and the advance toward Tunisia, the last part of North Africa held by the Axis forces until they were thrown out in May 1943.

The XXII TAC served in the combat role within the Mediterranean theatre until the end of World War II, and was then deactivated at Pomigliano, Italy, on 4 October 1945.

Known components of the XXII TAC were a pair of wings flying the P-47 Thunderbolt single-engine heavy fighter. These units were the 63rd Fighter Wing comprising the 52nd Fighter Group (1943–44) and the 350th Fighter Group (1943–44), and the 87th Fighter Wing comprising the 57th Fighter Group (1944), 79th Fighter Group (1944) and 86th Fighter Group (1944).

▲ Heavy cruiser

The P-47N was the fastest and heaviest of the Thunderbolts to fly in World War II, and included square-cut wing tips and extra fuel cells in the root of each wing.

Thunderbolt flyers

Between them, these five groups mustered a total of 15 squadrons with some 240 aircraft. In a theatre essentially bereft of German fighter opposition and bombers that could otherwise have been the targets of their heavy fixed forward-firing armament of six or eight 12.7mm (0.5in) machine guns, the Thunderbolts were operated almost exclusively in the fighter-bomber role. This role was especially important in the Italian theatre, as artillery was more limited in its applications than it would otherwise have been in less rugged terrain.

The Germans displayed skill and industry in creating excellent defensive complexes, which carefully exploited the advantages offered to the defence by the backbone-like Apennine mountains, together with the hills scoured into a myriad steep gorges by fast-flowing rivers extending west and east from them. The fighter-bomber was therefore amongst the major weapons of choice for the Allied air forces. With useful firepower and disposable loads, the single-engine fighter-bomber could operate in twisty valleys at low level, strafing convoys in the open and bombing defensive emplacements.

Specifications

Crew: 1
Powerplant: 1891kW (2535hp) Pratt & Whitney R-2800-59W Double Wasp
Maximum speed: 697km/h (433mph)
Range: 3060km (1900 miles) with drop tanks
Service ceiling: 12,495m (41,000ft)
Dimensions: span 12.42m (40ft 9in); length 11.02m (36ft 2in); height 4.47m (14ft 8in)
Weight: 7938kg (17,500lb) maximum takeoff
Armament: 8 x 12.7mm (0.5in) MGs in wings, plus provision for 1134kg (2500lb) external bombs or rockets

▲ Republic P-47D-25-RE Thunderbolt

527th Fighter Squadron / 86th Fighter Group, Pisa, Italy, 1944

Sturdy and well armed with heavy machine guns and disposable weapons under the wing, the P-47D was a classic heavy fighter-bomber, and extra range was provided by the drop tank.

1945 Fighter Wing

By 1945, the USAAF's fighter strength had been integrated into powerful fighter wings. Each of these comprised three fighter groups, which in turn consisted of three fighter squadrons. Each fighter squadron had four flights each of four aircraft in the form of two tactical pairs, each based on a two-aeroplane team (leader and his wingman). In overall terms, therefore, this gave the fighter wing a total of 144 aircraft.

Fighter Group 1, Squadron 1

Flight 1 **Flight 2**

Flight 3 **Flight 4**

Squadron 2

Flight 1 **Flight 2**

Flight 3 **Flight 4**

Squadron 3

Flight 1 **Flight 2**

Flight 3 **Flight 4**

Fighter Group 2, Squadron 1

Flight 1 **Flight 2**

Flight 3 **Flight 4**

Squadron 2

Flight 1 **Flight 2**

Flight 3 **Flight 4**

Squadron 3

Flight 1 **Flight 2**

Flight 3 **Flight 4**

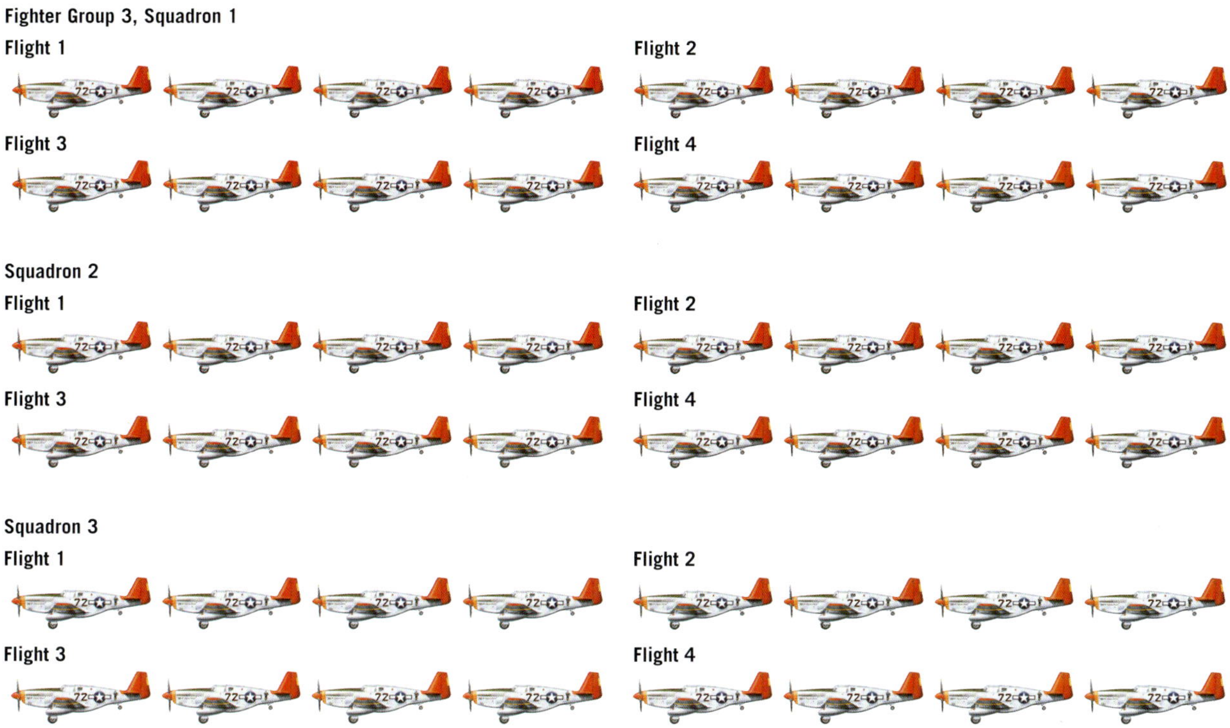

US Fifteenth Air Force

1943–45

The Fifteenth Air Force was the southern counterpart of the UK-based Eighth Air Force, and gave the United States the ability to undertake the strategic bomber role in Italy, the Balkans as far east as the Black Sea, and central European areas.

The US Fifteenth Air Force was one of the later additions to the overall command organization of the USAAF in World War II for service in the Mediterranean theatre. It was conceived as a strategic air force, with emphasis therefore placed on its heavy bomber capability, and it embarked on its programme of combat operations only one day after it had been established. This was made possible by the fact that the new air force inherited already well-established units, and the Fifteenth Air Force commander, Lieutenant-General 'Jimmy' Doolittle, was well-experienced in the organization and control of strategic bomber operations and was thus in the position to press ahead without delay.

ORGANIZATION

US Fifteenth Air Force
August 1944

- 5th Bombardment Wing
- 47th Bombardment Wing
  - 304th Bombardment Wing
  - 55th Bombardment Wing
- 49th Bombardment Wing
- 306th Fighter Wing

The new air force force drew its operational elements from the heavy bomber units of the IX Bomber Command of the Ninth Air Force, which was being relocated to the UK to become the

USAAF's primary exponent of tactical air power in the north-west Europe campaign, the XII Bomber Command of the Twelfth Air Force, and diversion of groups for the Eighth Air Force.

Strategic role

Operating mainly from bases in southern Italy, the Fifteenth Air Force, along with the Eighth Air Force and RAF Bomber Command, became the Allied instruments of the strategic air war waged against Germany. The Fifteenth Air Force lost 2110 bombers from its 15 B-24 and six B-17 bombardment groups, while its seven fighter groups claimed a total of 1836 German aircraft destroyed. The Fifteenth Air Force was deactivated on 15 September 1945.

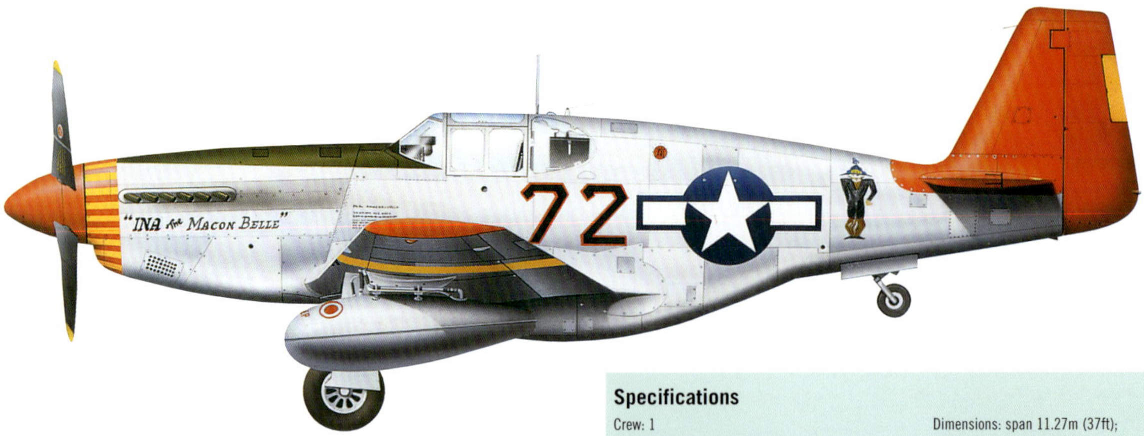

Specifications

Crew: 1
Powerplant: 1044kW (1400hp) Packard V-1650-3 Rolls-Royce Merlin engine
Maximum speed: 690km/h (430mph)
Range: 3540km (2200 miles)
Service ceiling: 12,649m (41,500ft)
Dimensions: span 11.27m (37ft); length 9.84m (32ft 4in); height 4.15m (13ft 8in)
Weight: 4173kg (9200lb) loaded
Armament: 6 x 12.7mm (0.5in) MGs

▲ **North American P-51C Mustang**

302nd Fighter Squadron / 322nd Fighter Group, Italy

This was the aeroplane of the ace Lieutenant Lee 'Buddy' Archer.

Specifications

Crew: 1
Powerplant: 1044kW (1400hp) Packard V-1650-3 Rolls-Royce Merlin engine
Maximum speed: 690km/h (430mph)
Range: 3540km (2200 miles)
Service ceiling: 12,649m (41,500ft)
Dimensions: span 11.27m (37ft); length 9.84m (32ft 4in); height 4.15m (13ft 8in)
Weight: 4173kg (9200lb) loaded
Armament: 6 x 12.7mm (0.5in) MGs

▲ **North American P-51B Mustang**

318th Fighter Squadron / Fifteenth Air Force

The one major limitation of the otherwise excellent P-51B was its framed canopy, whose large areas of metal impaired the pilot's lateral fields of vision.

Italy and Southern Europe
1943–45

The aircraft of the US Fifteenth Air Force, operating primarily from bases in the flat areas near the 'heel' of Italy, played a major part in curtailing the flow of supplies to Germans forces seeking desperately to hold onto the northern half of Italy from the summer of 1944.

As the two Allied armies, the US Fifth Army in the west and the British Eighth Army in the east, advanced north from the line of Rome, which fell on 4 June 1944, they continued to encounter the standard pattern of skilled German defence.

Tactical air power was of vital importance in winkling the Germans from their defences through the use of light bombers and ground-attack aircraft, while medium bombers attacked farther to the north in an effort to cut the Germans' lines of communication and destroy rear-area facilities.

Targeting strategic facilities

There was also scope for heavy bombers to roam still deeper behind the front to attack strategic targets in the Balkans and the southern part of central Europe, including aircraft production facilities, tank manufacturing factories, and key industrial facilities and fuel production centres all across the region.

There was also the opportunity to disrupt, if not actually to sever, the limited number of routes that the Germans could use to nourish their ground forces in Italy. These rail and road routes included natural chokepoints such as the Brenner Pass, and also included major marshalling yards in cities such as Vienna and Ljubljana.

The same type of effort was also waged against the German communications to their forces in Greece, Albania and southern Yugoslavia. Little praised at the time or later, this strategic bombing effort played a major part in crippling the Germans.

ORGANIZATION

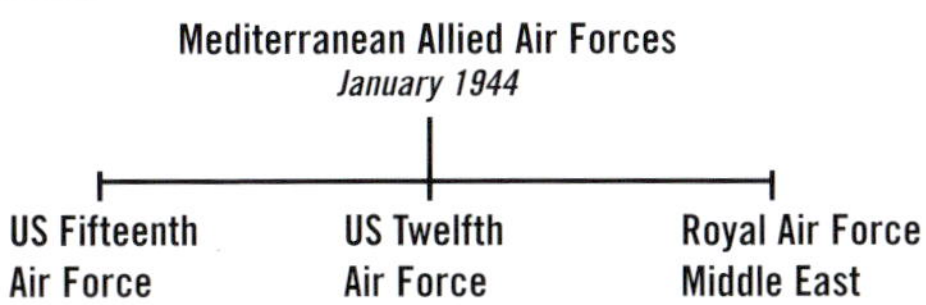

Fighter ace

Captain Andrew Turner (left), commanding officer of the 302nd Fighter Squadron, 332nd Fighter Group, and Lieutenant Clarence 'Lucky' Lester (right) discuss a recent mission.

306th Fighter Wing

1943–45

Like all other strategic air forces of the USAAF, the Fifteenth Air Force in Italy had the heavy bomber at its core, but also needed escort fighters. This capability was provided by the 25 fighter groups controlled by the 306th Fighter Group.

THE 306TH FIGHTER Group was the escort fighter element of the US Fifteenth Air Force, and operated two types of fighter in the form of the Lockheed P-38 Lightning twin-engine warplane and the North American P-51 Mustang single-engine fighter. The three Lightning elements were the 1st, 14th and 82nd Fighter Groups, which were based at Salsola, Triolo and Vincenzo respectively.

The 1st Fighter Group, which moved to Aghione in Corsica for a time in August 1944 to support the *Dragoon* landings in southern France, comprised the 27th, 71st and 94th Fighter Squadrons; the 14th Fighter Group comprised the 37th, 37th and 48th Fighter Squadrons; and the 82nd Fighter Group comprised the 95th, 96th and 97th Fighter Squadrons.

▲ **Mountain cruising**

'Bubbletop' Mustangs of the 325th (foreground), 332nd, 52nd and 31st Fighter Groups pose for the camera late in the war. Note the lack of colourful markings on the Mustang from the 332nd Fighter Group.

Mustang flyers

The four Mustang elements were the 31st, 52nd, 325th and 332nd Fighter Groups, which were based at San Severo, Madna, Lesina and Ramitelli respectively. The 31st Fighter Group comprised the 307th, 308th and 309th Fighter Squadrons; the 52nd Fighter Group comprised the 2nd, 4th and 5th Fighter Squadrons; the 325th Fighter Group comprised the 317th, 318th and 319th Fighter Squadrons; and the 332nd Fighter Group comprised the 99th, 100th, 301st and 302nd Fighter Squadrons, although it should be noted that the 99th Fighter Squadron was allocated to the 86th Fighter Group of the US Twelfth Air Force, and flew the Curtiss P-40 Warhawk and the Republic P-47 Thunderbolt as well as the Mustang.

306TH FIGHTER WING

Base	Group	Squadrons
Salsola	1st FG	27th, 71st & 94th FSs
Triolo	14th FG	37th, 47th & 48th FSs
San Severo	31st FG	307th, 308th & 309th FSs
Madna	52nd FG	2nd, 4th & 5th FSs
Vincenzo	82nd FG	95th, 96th & 97th FSs
Lesina	325th FG	317th, 318th & 319th FSs
Ramitelli	332nd FG	99th, 100th, 301st & 302nd FSs

Free roaming role

While the primary task of these fighters units was the escort of the Fifteenth Air Force's B-24 Liberator heavy bombers, the decline of the *Luftwaffe*'s fighter arm, whose remnants were increasingly concentrated for the defence of Germany, meant that the Allied fighters had what was, in effect, a roving commission to search out German air power at low as well and medium and high altitudes, and in the absence of aerial 'trade' to search for targets of opportunity such as convoys, trains, fuel storage facilities and anything resembling a military dump.

ORGANIZATION

306th Fighter Wing
spring 1945

- 1st Fighter Group
- 14th Fighter Group
- 31st Fighter Group
- 52nd Fighter Group
- 82nd Fighter Group
- 325th Fighter Group
- 332nd Fighter Group

▲ North American P-51D Mustang

307th Fighter Squadron / 31st Fighter Group, Italy, 1944–45

This was the aeroplane of Lieutenant James Brooks, a pilot of the US Fifteenth Air Force.

Specifications

Crew: 1
Powerplant: 1264kW (1695hp) Packard Merlin V-1650-7 V-12 piston engine
Maximum speed: 703km/h (437mph)
Range: 3347km (2080 miles)
Service ceiling: 12,770m (41,900ft)
Dimensions: span 11.28m (37ft); length 9.83m (32ft 3in); height 4.17m (13ft 8in)
Weight: 5488kg (12,100lb) loaded
Armament: 6 x 12.7mm (0.5in) MGs in wings, plus up to 2 x 454kg (1000lb) bombs

Specifications

Crew: 1
Powerplant: 1264kW (1695hp) Packard Merlin V-1650-7 V-12 piston engine
Maximum speed: 703km/h (437mph)
Range: 3347km (2080 miles)
Service ceiling: 12,770m (41,900ft)
Dimensions: span 11.28m (37ft); length 9.83m (32ft 3in); height 4.17m (13ft 8in)
Weight: 5488kg (12,100lb) loaded
Armament: 6 x 12.7mm (0.5in) MGs in wings, plus up to 2 x 454kg (1000lb) bombs

▲ North American P-51D Mustang

317th Fighter Squadron / 325th Fighter Group, Italy 1944–45

This was the aeroplane of Lieutenant Walter R. Hinton, a pilot of the US Fifteenth Air Force. Note the diagonal-chequered pattern on the tailplane, used by all aircraft throughout the 325th Fighter Group.

Chapter 4

Soviet Union

Although the Soviet air forces were numerically strong in 1939, they were in a poor state of training and the vast majority of their aircraft were obsolete. The 'Winter War' of 1939–40 with Finland proved this point, and the Soviet Union accelerated the modernization programme it had already begun on the basis of some advanced warplanes under development. This process was wholly incomplete at the time of the German invasion of June 1941, but the destruction of vast numbers of Soviet aircraft was something of a blessing in disguise. A new Soviet air arm began to emerge on the basis of good warplanes optimized for Soviet production and operating conditions, and a training programme that emphasized skills rather than an adherence to set procedures. From 1943, the Soviets steadily swept the Germans from the skies.

On patrol

Two Polikarpov I-16 fighters patrol over the vast forests of northern Russia before the outbreak of war with Nazi Germany.

Russo-Finnish War
1939–40

When he ordered the invasion of Finland in November 1939, Stalin believed that the Soviet forces would secure quick and total victory. Despite being totally outnumbered, in the air as on the ground, the Finns resisted with great determination and there followed a bloody campaign.

THE FINNS POSSESSED 145 frontline aircraft, of which only 114 were serviceable, while the Soviets could call on up to 3800 aircraft. After early reverses, the Soviets sent in larger numbers of more modern aircraft, but the Finns were for some time able to cope with this as a result of their superior tactics. Finland used the 'finger four' formation – four aircraft flying as two pairs, one low and the other high, each fighting independently but supporting its wingman in combat. This was better than the Soviet tactic of three fighters in a 'vic' formation. The Finnish tactics, combined with the Finnish belief in attack rather than defence, regardless of the odds, contributed to the failure of Soviet bombers to inflict substantial damage on Finnish positions, industries and population centres.

Specifications
Crew: 1
Powerplant: 358kW (480hp) M-22 radial piston engine
Maximum speed: 278km/h (173mph)
Range: 560km (348 miles)
Service ceiling: 7500m (24,605ft)
Dimensions: span 10.24m (33ft 7in); length 6.78m (22ft 3in)
Weight: 934kg (2059lb) unloaded
Armament: 2 x 7.62mm (03in) PV-1 MGs, plus 2 x 20kg (44lb) bombs

▲ **Polikarpov I-15**
Unknown unit, Soviet Union, late 1930s
This aeroplane has the standard dark-green summer camouflage of the period.

Specifications
Crew: 1
Powerplant: 746kW (1000hp) M-62 radial piston engine
Maximum speed: 490km/h (304mph)
Range: 600km (373 miles)
Service ceiling: 9470m (31,070ft)
Dimensions: span 8.88m (29ft 2in); length 6.04m (19ft 10in)
Weight: 2060kg (4542lb) loaded
Armament: 4 x 7.62mm (03in) ShKAS MGs (sometimes 2 x 20mm/0.8in replacing 2 x ShKAS MGs in wings); up to 200kg (441lb) bomb load

▲ **Polikarpov I-16 Type 10**
Unknown unit, Soviet Union, 1939
This version is depicted in the standard green summer camouflage.

Specifications

Crew: 1

Powerplant: 578kW (775hp) M-25V radial piston engine

Maximum speed: 370km/h (230mph)

Range: 560km (348 miles)

Service ceiling: 9500m (31,070ft)

Dimensions: span 10.20m (33ft 6in); length 6.27m (20ft 7in)

Weight: 1900kg (4189lb) loaded

Armament: 4 x 7.62mm (03in) ShKAS MGs; up to 150kg (331lb) bomb load

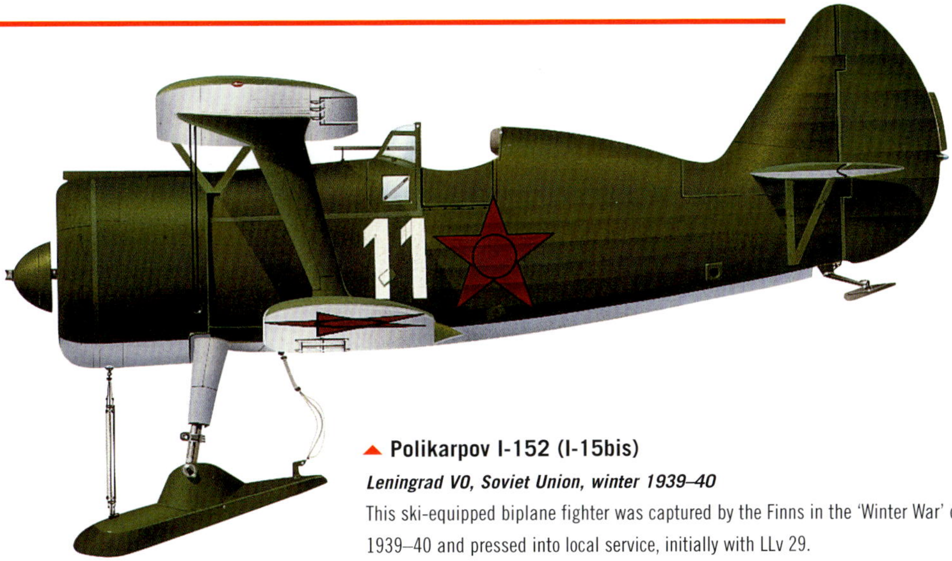

▲ **Polikarpov I-152 (I-15bis)**

Leningrad VO, Soviet Union, winter 1939–40

This ski-equipped biplane fighter was captured by the Finns in the 'Winter War' of 1939–40 and pressed into local service, initially with LLv 29.

Soviet V-VS

1941–42

When the Germans unleashed Operation *Barbarossa,* the invasion of the Soviet Union in June 1941, the Soviet air forces were large but technically obsolete and in a dismal state of training that emphasized orthodoxy rather than initiative. Change was under way, but it was too late to prevent catastrophic losses during 1941.

THE SOVIET AIR FORCE, which is also known by the abbreviation V-VS (*Voenno-Vozdushnye Sily*), was the official designation of one of the air forces of the Soviet Union, the other being the Soviet Anti-Air Defence (PVO). The origins of the V-VS can be found in the the so-called Workers' and Peasants' Air Fleet established in 1918 as the Soviet successor to the Imperial Russian Air Force.

After the service was placed under control of the Red Army, which led to the change of designation to V-VS in 1930, the air force was much less influenced by the thinking and planning of the Communist party and more geared towards serving the needs of the professional Red Army.

After the creation of the Soviet Union, many efforts were made both to modernize the Soviet aero industry on a national basis and to enlarge aircraft production. As a result, the production of aircraft increased rapidly and dramatically in the USSR in the early years of the 1930s. In the middle and late years of that decade, the V-VS could therefore introduce Polikarpov I-15 biplane and Polikarpov I-16 monoplane fighters, the former including a variant with retractable main landing gear units, and also the Tupolev SB-2 and Ilyushin DB-3 monoplane bombers, the former generally acknowledged as the best high-speed bomber of its time.

ORGANIZATION

V-VS
May 1942

First Air Army | Second Air Army | Third Air Army | Fourth Air Army | Fifth Air Army

Early test

One of the first real tests of the V-VS and the Soviet aero industry came in 1936 with the start of the Spanish Civil War. The Soviet Union decided to aid the Republican government side with equipment and some manpower, and here the best of the V-VS was put to the test against German and Italian detachments flying for the Nationalist rebel side and

SOVIET FIGHTER TYPES JUNE 1941			
Type	**Speed**	**Armament**	**Number built**
Lavochkin LaGG-3	560km/h (348mph)	1 x 20mm cannon 2 x 12.7mm MGs	6527
Mikoyan MiG-3	640km/h (398mph)	1 x 12.7mm 2 x 7.62mm MGs	3322
Polikarpov I-15bis	370km/h (230mph)	4 x 7.62mm MGs	2408
Polikarpov I-153	365km/h (227mph)	4 x 7.62mm MGs	3437
Polikarpov I-16	525km/h (326mph)	2 x 20mm cannon 2 x 7.62mm MGs	c.6500
Yakovlev Yak-1	580 km/h (360mph)	1 x 20mm cannon 2 x 7.62mm MGs	8720

testing their own latest types. Early success with the I-16 was wasted because of the limited use of this fighter, and the arrival of the Messerschmitt Bf 109 later in the war secured air superiority for the Nationalists. At the start of World War II, the Soviet military was not yet ready for developing modern technology suitable for winning a war.

In 1939, the V-VS used its bombers to attack Finland in the 'Winter War', but the losses inflicted on it by the relatively small Finnish air arm showed the air force's shortcomings, mostly due to loss of personnel in the purges of the late 1930s.

Specifications

Crew: 1
Powerplant: 820kW (1100hp) M-63 radial piston engine
Maximum speed: 2490km/h (304mph)
Range: 600km (373 miles)
Service ceiling: 9470m (31,070ft)
Dimensions: span 8.88m (29ft 2in); length 6.04m (19ft 10in)
Weight: 2060kg (4542lb) loaded
Armament: 4 x 7.62mm (03in) ShKAS MGs (sometimes 2 x 20mm [0.8in] replacing 2 x ShKAS MGs in wings); up to 200kg (441lb) bomb load

Polikarpov I-16 Type 24

Unknown squadron of the Leningrad Military District, Soviet Union, summer 1939

The I-16 was very tricky to fly as a result of its short and tubby fuselage, and nearly all pilots preferred an open cockpit.

Barbarossa: the first months

June–August 1941

In the first three months of their campaign against the Soviet Union, the Germans swept all before them. The V-VS lost many thousands of aircraft and men, but already manufacture of more modern aircraft was being stepped up, and training was improved.

The German invasion of the Soviet Union started on 22 June 1941. Three-fifths of the *Luftwaffe*'s strength was deployed along the frontier with the Soviet Union. Some 1400 of the 1945 operational aircraft, 1280 of them serviceable, were gathered in four *Luftflotten:* 650 fighters, 831 bombers, 324 dive-bombers and 140 reconnaissance aircraft, as well as 200 transports plus coastal and reconnaissance machines. The Germans had allied support, including 299 Finnish aircraft just a short time later,

Specifications

Crew: 1
Powerplant: 954kW (1280hp) Rolls-Royce Merlin XX liquid-cooled V-12
Maximum speed: 529km/h (329mph)
Range: 1480km (920 miles)
Service ceiling: 10,850m (35,600ft)
Dimensions: span 12.19m (40ft 0in); length 9.81m (32ft 2.25in); height 3.98m (13ft 1in)
Weight: 3629kg (8044lb) loaded
Armament: 8 x 7.7mm (0.303in) Browning MGs

▲ **Hawker Hurricane Mk IIB**

No. 18 Squadron / No. 151 Wing / RAF, Vaenga, autumn 1941

In an effort to show solidarity with the Soviet people after the German invasion of the Soviet Union in June 1941, the British sent a wing of Hurricane fighters, complete with air and ground crews, in the late summer of 1941. The aircraft flew with Soviet markings superimposed on the British camouflage.

ORGANIZATION

V-VS Disposition
June 1941

- **Baltic Military District (630 aircraft)**
- **Western Military District (1630 aircraft)**
- **Kiev Military District (1085 aircraft)**
- **Leningrad Military District (1155 aircraft in 24 regiments of eight divisions)**
- **Odessa Military District (part of 1400 aircraft shared with other districts)**

but the Axis air strength was wholly overshadowed in numerical terms by the V-VS' strength of some 12,000–15,000 operational aircraft, about 7000 of them in the west, in 23 air divisions.

The Soviets were currently working their way through a great programme of re-equipment and retraining. Not expecting a German attack, they were also based in undispersed form on airfields right up against the frontier. However, the *Luftwaffe* possessed a huge advantage in terms of experience, and also in terms of the superior quality of the aircraft available to them.

Surprise attack

The first attack was made just after 3 a.m. by 637 bombers and 331 fighters against 31 airfields. The Soviets were caught completely by surprise, and the Germans lost a mere two aircraft. There followed a stream of later raids, and by the end of the day 1489 Soviet aircraft had been claimed on the ground and 322 in the air as the German ground forces swept forward.

The Germans' standard tactic was the employment of dive-bombers and heavy fighters in direct support of the armoured spearheads, these aircraft acting as airborne artillery while medium bombers attacked supply and concentration areas, railway communications, convoys and targets deeper behind the front. In the first seven days, the Soviets lost 4990 aircraft and the Germans just 179.

The greatest battles of the war up to this time took place as the German spearheads approached Minsk on 30 June. Large formations of unescorted Soviet bombers were committed, and one German unit, *Jagdgeschwader* 51, claimed 114 of these shot down. Further north, while defending bridges captured over the Duna river, JG 54 claimed the destruction of 65 bombers shot down, and on 6 July some 65 of 73 bombers attacking a German bridgehead at Ostrov were destroyed.

The first air raid on Moscow was flown on 22 July: 127 bombers dropped 106 tonnes (104 tons) of bombs. On the following night, 115 bombers attacked, and 100 more went back on the day after this. Thereafter Moscow was bombed regularly by day as well as night. During the summer months, the superior German fighters were notably active, and the unfortunate Soviet aircraft were shot down in almost unbelievable numbers.

The defence of Moscow

November 1941 – February 1942

The Germans had gambled on winning a swift victory in the Soviet Union, but were then checked in front of Moscow by renewed Soviet strength and the onset of truly bitter winter conditions. Then the Soviets counterattacked, driving the Germans back.

During September 1941, the Soviets had attempted several counterattacks on the German Army Group Centre directed towards Moscow but, now woefully short of armoured vehicles as well as aircraft, they could not attempt a strategic counteroffensive. On 2 October, the Germans committed Army Group Centre to a final push on the capital. In front of this formation was a complex of defence lines, but the Germans broke through the first of these without difficulty and took Orel some 120km (75 miles) behind it. The Germans then closed in and trapped huge numbers of Soviet personnel, leaving the Soviets with just 90,000 men and 150 tanks for the defence for Moscow. But the weather was worsening steadily, and the Germans paused to regroup, which allowed the Soviets to bring in fresh troops from Siberia. The Germans got under way once more on 15 November, by 2 December reaching a point only 24km (15 miles) west of central Moscow. But the first blizzards of the winter were now beginning, and the Germans forces lacked the equipment for winter survival, let alone combat. The Soviets now counterattacked and by February 1942 had pushed the Germans back.

GERMAN & SOVIET FRONTLINE AIR STRENGTHS

Date	German	Soviet
June 1941	2130	8100
July 1941	1050	2500
December 1941	2500	2500
May 1942	3400	3160

ORGANIZATION

Air Defence of Moscow
July 1941

- Western Frontal Aviation
- Reserve Frontal Aviation
- Bryansk Frontal Aviation
- VI Fighter Air Corps

▲ **Mikoyan-Gurevich MiG-3P**

6th IAP (Fighter Aviation Regiment) / 6th IAK (Fighter Aviation Corps) / PVO / Moscow Air Defence Zone, early 1942

This aeroplane was flown by A. V. Shlopov.

Specifications

Crew: 1
Powerplant: 1007kW (1350hp) Mikuli AM-35A V12 piston engine
Maximum speed: 640km/h (398mph)
Range: 1195km (743 miles)
Service ceiling: 12,000m (39,370ft)
Dimensions: span 10.20m (33ft 6in); length 8.26m (27ft 1in); height 3.50m (11ft 6in)
Weight: 3350kg (7385lb) loaded
Armament: 1 x 12.7mm (0.5in) Beresin and 2 x 7.62mm (0.3in) ShKAS MGs; up to 200kg (441lb) bomb load or 6 x RS-82 rockets on underwing racks

Specifications

Crew: 1
Powerplant: 1007kW (1350hp) Mikuli AM-35A V12 piston engine
Maximum speed: 640km/h (398mph)
Range: 1195km (743 miles)
Service ceiling: 12,000m (39,370ft)
Dimensions: span 10.20m (33ft 6in); length 8.26m (27ft 1in); height 3.50m (11ft 6in)
Weight: 3350kg (7385lb) loaded
Armament: 1 x 12.7mm (0.5in) Beresin and 2 x 7.62mm (0.3in) ShKAS MGs; up to 200kg (441lb) bomb load or 6 x RS-82 rockets on underwing racks

▲ **Mikoyan-Gurevich MiG-3**

34th IAP / 6th IAK, Moscow, winter 1941–42

'Za Stalina' (For Stalin) was a common legend on Soviet aircraft of the period.

Specifications

Crew: 1
Powerplant: 925kW (1240hp) Klimov M-105PF V12 piston engine
Maximum speed: 560km/h (348mph)
Range: 650km (404 miles)
Service ceiling: 9600m (31,490ft)
Dimensions: span 9.80m (32ft 2in); length 8.90m (29ft 3in)
Weight: 3280kg (7231lb) loaded
Armament: 2 x 20mm (0.8in) cannon (firing through propeller hub), 2 x 12.7mm ((0.5in) MGs in nose

▲ **Lavochkin LaGG-3**

6th IAK / Moscow Air Defence Zone, summer 1942

The unusual black and green camouflage was the result of the aeroplane being built in a factory that had previously manufactured tractors, which had always been completed in black or green.

The defence of Leningrad

August 1941 – February 1942

Named for the first leader of the Soviet Union, Leningrad had a huge symbolic attraction to Hitler. The Germans closed on the city in August 1941, but the Soviets were wholly determined to hold the city, no matter how heavy the cost.

On 1 September 1941, the forces of the German Army Group North were close enough to the great city of Leningrad to start their artillery bombardment. The Germans, who had decided against a direct assault on the city, now began the process of encircling it with the aid of the Finnish Army in the north, aiming to starve the city's garrison and civil population and so force them to surrender.

On 15 September, with the capture of Schlusselburg on the southern edge of Lake Ladoga, the city was completely cut off from overland communication with the rest of the Soviet Union. The food remaining in the city was sufficient, if carefully rationed, for about one month, and people began to succumb to starvation during October and November, some 11,000 deaths recorded in the latter month. A supply route from Tikhvin to Lednevo had been used to get some food across the lake in barges, but on 9 November the Germans took Tikhvin, and in the middle of the month the ice on the lake made navigation impossible while not yet being thick enough to support heavy motor vehicles such as laden trucks.

A new road from Zaboriye via Karpino to Lednevo was opened on 6 December, but the combination of the poor surface conditions and deteriorating weather meant that trucks could achieve little more than 30km (19 miles) per day on it. On 9 December, as part of the Red Army's winter counteroffensive, Tikhvin was recaptured and the Germans were pushed back to the Volkhov river. The railway was repaired and an ice road opened across the lake, facilitating a proper supply line to the city. But the supply situation remained precarious; on 25 December alone, 3700 people died. The situation became more desperate in the spring of 1942, when the thaw of the ice made the survival of Leningrad still more problematical.

Fighter regiment, 1943

In 1943, the Soviet fighter arm was based on the fighter regiment, itself subordinated to the fighter division (three regiments) or the composite division (one fighter, one bomber and one ground-attack regiments). In basic terms the regiment comprised three fighter squadrons, each with about 12 aircraft, each squadron being subdivided into three sections of four aircraft. Each squadron tended to be equipped with the same type of aircraft, but a regiment might include a mixture of types, as shown here.

1st Squadron (12 x MiG-3s)

2nd Squadron (12 x Yak-1s)

3rd Squadron (12 x Yak 1-Ms)

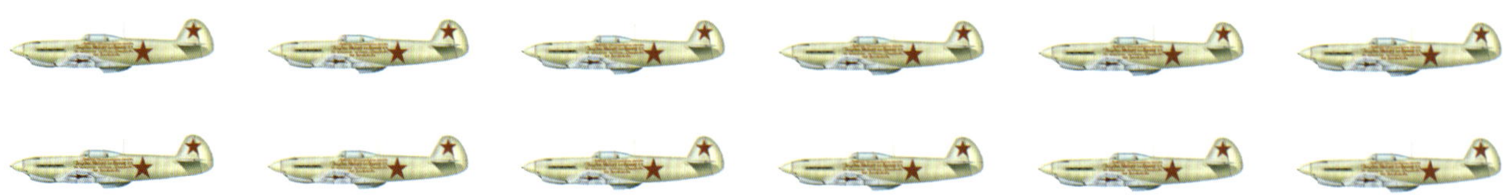

Soviet V-VS
1942–45

During 1942, the V-VS began to recover strongly from its decimation in 1941, with new aircraft and better training. Soviet production also grew in 1942, and by 1943 the fighter arm of the V-VS was in a position to start challenging and then getting the better of the *Luftwaffe*.

As it started to recover from the disasters of 1941, the V-VS was able to rebuild, largely from the efforts of factories that had been uprooted from European Russia and re-erected to the east of the Ural mountain barrier, an area in which other factories also sprang up. Initially the V-VS received the same types of aircraft it had been operating in 1941, for which the manufacturing skills already existed and which made large-scale use of materials readily available in the Soviet Union, most notably two forms of plywood created from local wood. The same materials also featured in the new types of aircraft that began to emerge from the production lines in 1942. So far as fighters were concerned, these were primarily the products of the Lavochkin, Mikoyan-Gurevich and Yakovlev design organizations. The two most important were the first and third, which created basic designs of advanced concept that could then be built in large numbers and steadily enhanced by incremental improvement.

Fighter power

The Soviets' main strength was always its fighters, for which emphasis was placed on simple manufacture and maintenance, strength to absorb battle damage and the rigours of operating on primitive airfields,

▲ **Mikoyan-Gurevich MiG-3 squadron**

Here, pilots of the 12th Soviet Fighter Regiment take the 'oath of the guard', following combat successes in the winter of 1942.

ORGANIZATION

New Soviet Air Armies
June–December 1942

- Sixth Air Army
- Seventh Air Army
- Eighth Air Army
- Ninth Air Army
- Tenth Air Army
- Eleventh Air Army
- Twelfth Air Army
- Thirteenth Air Army
- Fourteenth Air Army
- Fifteenth Air Army
- Sixteenth Air Army
- Seventeenth Air Army

▲ Bell P-39Q Airacobra

16th Guards IAP, Eastern Front, 1944

This was one of the P-39 aircraft flown by Aleksandr Pokryshkin, the second highest-scoring Soviet ace, who shot down 48 of his eventual 59 'kills' while flying the Airacobra. Pokryshkin's memoirs and personal notebooks put his score at over 100. However, the Soviet air force did not officially confirm kills whose wreckage could not be found, and so many aircraft shot down over enemy territory were never confirmed.

Specifications

Crew: 1
Powerplant: 895kW (1200hp) Allison V-1710-85 liquid-cooled V-12
Maximum speed: 605km/h (376mph)
Range: 483km (300 miles)
Service ceiling: 11,665m (38,270ft)
Dimensions: span 10.36m (34ft 0in); length 9.195m (30ft 2in); height 3.785m (12ft 5in)
Weight: 3992kg (8,800lb) loaded
Armament: 1 x 37mm (1.46in) M4 cannon, 2 x 12.7mm (0.5in) Browning MGs, 4 x 7.62mm (0.3in) Browning MGs

▲ Lavochkin La-5FN

240th IAP, Eastern Front, April 1944

This aeroplane was flown by Ivan Kozhedub, a Hero of the Soviet Union and the highest-scoring Soviet ace of World War II. The legend below the cockpit translates as 'from collective farm worker Konev, Vasili Viktorovich'.

Specifications

Crew: 1
Powerplant: 1380kW (1850hp) Shvetsov M-82FN radial piston engine
Maximum speed: 665km/h (413mph)
Range: 635km (395 miles)
Service ceiling: 10.800m (35,435ft)
Dimensions: span 9.80m (32ft 2in); length 8.60m (28ft 3in)
Weight: 3280kg (7231lb) loaded
Armament: 2 x 20mm (0.8in) Beresin B-20 cannon in nose; plus up to 200kg (441lb) bomb load

good performance and handling, powerful armament, and engines optimized for performance at up to 5000m (16,400ft). The qualities of these aircraft resulted from the decision made in the 1930s that the Soviet air forces would concentrate on tactical air warfare above the front in support of the ground forces, which were seen as the decisive arm. By 1943, therefore, the V-VS was considerably more of a match for the *Luftwaffe*, and in 1944 and 1945 achieved almost total air superiority.

GERMAN & SOVIET FRONTLINE AIR STRENGTHS

Date	German	Soviet
November 1942	2450	3100
July 1943	2500	8300
January 1944	1800	8500
June 1944	1710	11,800
January 1945	1430	14,500
April 1945	1500 (all fronts)	17,000

Specifications

Crew: 1
Powerplant: 880kW (1180hp) Klimov M-105PF V-12 liquid-cooled engine
Maximum speed: 592km/h (368mph)
Range: 700km (435 miles)
Service ceiling: 10,050m (33,000ft)
Dimensions: span 10 m (32ft 10in); length 8.5m (27ft 11in)
Weight: 2883kg (6343lb) loaded
Armament: 1 x 20mm (0.8in) ShVAK cannon, 1 x 12.7mm (0.5in) Berezin UBS MG

▲ **Yakovlev Yak-1M**

1st 'Warszawa' IAP, Eastern Front, late 1944

This squadron was manned by Polish units in the Soviet army. A single one-second burst from the Yak-1s combined machine guns using high-explosive ammunition produced a highly destructive 2kg (4.4lb) weight of fire.

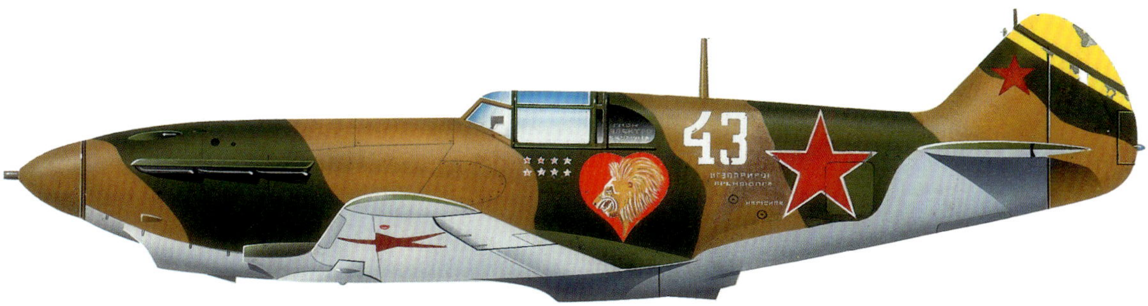

Specifications

Crew: 1
Powerplant: 925kW (1240hp) Klimov M-105PF V12 piston engine
Maximum speed: 560km/h (348mph)
Range: 650km (404 miles)
Service ceiling: 9600m (31,490ft)
Dimensions: span 9.80m (32ft 2in); length 8.90m (29ft 3in)
Weight: 3280kg (7231lb) loaded
Armament: 2 x 20mm (0.8in) cannon (firing through propeller hub), 2 x 12.7mm (0.5in) MGs in nose

▲ **Lavochkin LaGG-3**

9th IAP / V-VS ChF (Black Sea Air Force), southern Soviet Union, May 1944

Flown by Yuri Shchipov, this aeroplane carried the pilot's own 'lion's head' marking and eight victory symbols below the cockpit.

The Eastern Front
1942–43

The Soviets stunned the Germans with their counteroffensives late in 1941, and pressed their advantage into the spring thaw of 1942. Then the Germans replied with their own offensives, taking them towards the Caucasus and, fatally, the city of Stalingrad on the Volga river.

THE YEAR 1942 STARTED with the continuation of the Soviet counteroffensive that had checked the Germans in front of Moscow, this counteroffensive stretching along the full length of the Eastern Front. It ended in stalemate only after the advent of the spring thaw. Germany planned to regain the strategic initiative with a summer offensive towards the Caspian Sea and the important oilfields of the Caucasus, and preliminary offensives between early May and late June paved the way for this effort. A subsidiary undertaking took Sevastopol on the western side of the Crimea.

ORGANIZATION

Air Army Deployments
May 1942

- First Air Army (West Front)
- Second Air Army (South-West Front)
- Third Air Army (Kalinin Front)
- Fourth Air Army (South Front)
- Fifth Air Army (North Caucasus Front)

The German offensive, Operation Blue, began on 28 June and initially made good progress. Hitler then changed his mind about the objective, and added Stalingrad as one of the two major objectives. Divided aims meant divided forces, and Germany now lacked the strength to make fast progress on two diverging fronts. Even though the Soviet forces were in disarray, the determination of the Soviet high command, combined with the huge distances that the Germans were trying to cover, meant that the offensive slowed and then stalled. Through the summer, the *Luftwaffe* kept local air superiority, but the V-VS introduced advanced bombers and fighters operated by competent aircrews.

Order of the Red Banner

Overflying the Crimea, the lead aeroplane in this formation of Yakovlev Yak-9 fighters carries the Order of the Red Banner, indicating that all of these aircraft belong to a Guards unit.

Specifications

Crew: 1

Powerplant: 1380kW (1850hp) Shvetsov M-82FN radial piston engine

Maximum speed: 665km/h (413mph)

Range: 635km (395 miles)

Service ceiling: 10.800m (35,435ft)

Dimensions: span 9.80m (32ft 2in); length 8.60m (28ft 3in)

Weight: 3280kg (7231lb) loaded

Armament: 2 x 20mm (0.8in) Beresin B-20 cannon in nose; plus up to 200kg (441lb) bomb load

▲ Lavochkin La-5

Unknown unit, Eastern Front, 1943

This is an early example without the later variants' cut-down rear fuselage and revised canopy, which offered much improved all-round fields of vision.

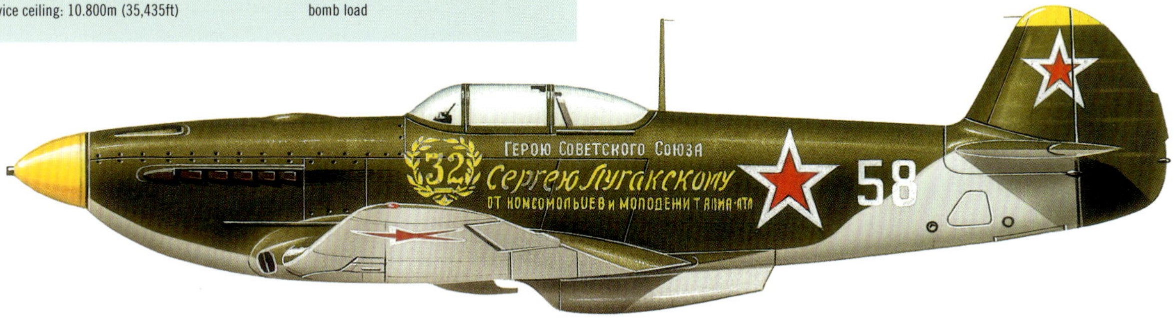

Specifications

Crew: 1

Powerplant: 962kW (1290hp) Klimov VK-105PF-2 V12 liquid-cooled piston engine

Maximum speed: 646km/h (401mph)

Range: 650km (405 miles)

Service ceiling: 10,700m (35,000ft)

Dimensions: span 9.2m (30 ft 2in); length 8.5m (27ft 11in); height 2.39m (7ft 11in)

Weight: 2692kg (5864lb) loaded

Armament: 1 x 20mm (0.8in) ShVAK cannon, 2 x 12.7mm (0.5in) Berezin UBS MG

▲ Yakovlev Yak-3

Unknown unit, Eastern Front, 1943

In common with many Soviet warplanes, this Yak-3 bears an exhortatory slogan on the side of the fuselage below the cockpit.

Specifications

Crew: 1

Powerplant: 880kW (1180hp) Klimov M-105PF V-12 liquid-cooled engine

Maximum speed: 592km/h (368mph)

Range: 700km (435 miles)

Service ceiling: 10,050m (33,000ft)

Dimensions: span 10 m (32ft 10in); length 8.5m (27ft 11in)

Weight: 2883kg (6343lb) loaded

Armament: 1 x 20mm (0.8in) ShVAK cannon, 1 x 12.7mm (0.5in) Berezin UBS MG

▲ Yakovlev Yak-1M

37th Guards IAP / 6th Guards AD, Eastern Front, 1943

This aeroplane was flown by Major B. N. Yevemen, and the legend under the cockpit translates as 'to the pilot of the Stalingrad Front Guards Major Comrade Yeven from the collective farm workers of the collective farm "Stakhanov", comrade F. P. Golovatov'.

Kursk

July 1943

The Battle of Kursk was the defining moment of the war on the Eastern Front. Here the Germans tried, for the last time, to regain the strategic initiative they had lost in the Battle of Stalingrad, and were decisively defeated by Soviet armoured, artillery and tactical air strength.

In February 1943, the German Sixth Army, trapped and starving in the ruins of Stalingrad, was forced to surrender its last 93,000 men after the *Luftwaffe* failed to deliver the tonnages of supplies it had promised to keep the beleaguered army in fighting trim. The Soviet forces launched a major offensive across the Donets river to take Kharkov, which the Germans then retook in an example of extraordinary combat and generalship skills. Both sides spent the period between March and July rebuilding their strengths, and planning for the summer. The lull in the fighting from March had left the Soviets with a large salient, centred on Kursk, bulging into the German lines beyond Orel in the north and Kharkov in the south.

Here the Germans saw an opportunity for a pincer offensive to strike south-east and north-east to meet behind Kursk, pinch out the salient and trap huge Soviet forces for annihilation. They gathered major armoured formations, rushed new armoured fighting vehicles into service and massed considerable air strength. But the Soviets were warned by their agents of the details of the German scheme, and prepared defences in great depth, with lines of earthworks, minefields, huge quantities of artillery and armour, and powerful air forces now able to meet the *Luftwaffe* in qualitative terms while exceeding it in quantitative terms.

ORGANIZATION

Soviet Air Armies in the Battle of Kursk
July 1943

Western Front (First Air Army with II Attack, II Fighter and VIII Fighter Corps)	Bryansk Front (Fifteenth Air Army with I Guards Fighter and III Attack Corps)
Central Front (Sixteenth Air Army with III Bomber, VI Combined and VI Fighter Corps)	Voronezh Front (Second Air Army with I Bomber, I Attack, IV Fighter and V Fighter Corps)

▲ Lavochkin La-5FN

Unknown unit, Eastern Front, 1943–44

The La-5 series of very capable fighters remained in production into 1944, 9920 such aircraft being delivered.

Specifications

Crew: 1
Powerplant: 1380kW (1850hp) Shvetsov M-82FN radial piston engine
Maximum speed: 665km/h (413mph)
Range: 635km (395 miles)
Service ceiling: 10,800m (35,435ft)
Dimensions: span 9.80m (32ft 2in); length 8.60m (28ft 3in)
Weight: 3280kg (7231lb) loaded
Armament: 2 x 20mm (0.8in) Beresin B-20 cannon in nose; plus up to 200kg (441lb) bomb load

The Battle of Kursk lasted between 5 and 16 July, and was a huge Soviet success. The northern half of the undertaking achieved little for the Germans, while their southern offensive managed to batter its way deeper into the Soviet lines before being checked. A major armoured battle, possibly the world's largest ever tank-versus-tank engagement, developed round Prokhorovka as the Germans tried to smash their way forward and the Soviets called in part of their huge reserve to defeat this effort.

Specifications

Crew: 1
Powerplant: 880kW (1180hp) Klimov M-105PF V-12 liquid-cooled engine
Maximum speed: 592km/h (368mph)
Range: 700km (435 miles)
Service ceiling: 10,050m (33,000ft)
Dimensions: span 10 m (32ft 10in); length 8.5m (27ft 11in)
Weight: 2883kg (6343lb) loaded
Armament: 1 x 20mm (0.8in) ShVAK cannon, 1 x 12.7mm (0.5in) Berezin UBS MG

▲ Yakovlev Yak-1

18th Guards IAP, Khationki, Eastern Front, spring 1943

This was the aeroplane of Lieutenant-Colonel A. E. Golubov. Golubov finished the war with 39 kills.

▲ Yakovlev Yak-9

Normandie-Niémen Squadron / French Air Force, Eastern Front, 1943–44

The *Normandie-Niémen* Squadron was formed from Free French fighter pilots at the suggestion of Charles de Gaulle, and served on the Eastern Front from March 1943 until the end of the war. During this time, the unit destroyed 273 enemy aircraft and received many orders, citations and decorations from both France and the Soviet Union.

Specifications

Crew: 1
Powerplant: 880kW (1180hp) Klimov M-105 PF V-12 liquid-cooled piston engine
Maximum speed: 591 km/h (367mph))
Range: 884km (549 miles)
Service ceiling: 9100m (30,000ft)
Dimensions: span 9.74m (31ft 11in); length 8.55m (28ft); height 3m (9ft 10in)
Weight: 3117kg (6,858lb) loaded
Armament: 1 x 20mm (0.8in) ShVAK cannon, 1 x 12.7mm (0.5in) Berezin UBS MG

NEW-GENERATION SOVIET FIGHTERS USED AT KURSK			
Type	Speed	Armament	Number built
Lavochkin La-5FN	650km/h (404mph)	2 x 20 or 23mm cannon	10,000
Yakovlev Yak-3	655km/h (407mph)	1 x 20mm cannon 2 x 12.7mm MGs	4848
Yakovlev Yak-7B	615km/h (382mph)	1x 20mm cannon 1 x 12.7mm MG	6399
Yakovlev Yak-9D	600km/h (372mph)	1 x 20mm cannon 1 x 12.7mm MG	16,769

Overhead the fighters of the V-VS prevented the *Luftwaffe* from playing a decisive part, and then supported the Soviet attack bombers, ground-attack aircraft and anti-tank warplanes in wreaking havoc on the German forces.

It is believed that the Germans lost something in the order of 3000 armoured fighting vehicles and 1400 aircraft at Kursk, and while their losses were equally as great the Soviets had decisively beaten the Germans. Hence between 12 July and 26 November the Soviets could unleash a number of great offensives between Smolensk and the Black Sea from which the *Wehrmacht* never recovered.

The push west

December 1943 – September 1944

Soviet forces swept west after Kursk, and the Germans were able to inflict no more than local reverses that merely delayed rather than halted the Soviet forces. Only after huge advances did logistical exhaustion finally bring the Red Army to a halt in the late summer of 1944.

Supported by ever larger numbers of tactical warplanes, such as the Ilyushin Il-2 ground-attack and Petlyakov Pe-2 light bomber types, and operating under an umbrella of ever more capable Lavochkin and Yakovlev fighters, the Soviet ground forces lifted the siege of Leningrad on 19 January 1944 even as the current winter offensive, launched late in the previous year, moved west.

▲ **Yakovlev Yak-9D**

Unknown unit, Eastern Front, 1944–45

The Yak-9D was the extended-range development of the basic Yak-9 fighter, with fuel capacity increased from 440l (115 US gal) to 650l (170 US gal), giving a maximum range of 1360km (845 miles).

Specifications

Crew: 1
Powerplant: 880kW (1180hp) Klimov M-105 PF V-12 liquid-cooled piston engine
Maximum speed: 591 km/h (367mph))
Range: 1360km (845 miles)
Service ceiling: 9100m (30,000ft)
Dimensions: span 9.74m (31ft 11in); length 8.55m (28ft); height 3m (9ft 10in)
Weight: 3117kg (6858lb) loaded
Armament: 1 x 20mm (0.8in) ShVAK cannon, 1 x 12.7mm (0.5in) Berezin UBS MG

Dive-bomber

As well as being a superb interceptor, the Yak-9 was also used in the ground-attack role.

Between January and April, the Soviet forces in the south crossed the great Dniepr river and liberated Ukraine before advancing into Romania, leading to the defection of that country and Bulgaria to the Soviet side in September.

With the launch of the Red Army's greatest ever ground offensive, Operation Bagration, in June and July, the Soviets liberated Belorussia in the Eastern Front's central sector, and exerted major pressure on

Specifications

Crew: 1
Powerplant: 1380kW (1850hp) Shvetsov M-82FN radial piston engine
Maximum speed: 680km/h (425mph)
Range: 990km (618 miles)
Service ceiling: 9500m (31,160ft)
Dimensions: span 9.80m (32ft 2in); length 8.60m (28ft 3in)
Weight: 3280kg (7231lb) loaded
Armament: 2 x 20mm (0.8in) ShVAK cannon or 3 x 20mm (0.8in) Berezin B-20 cannon in nose; plus up to 200kg (441lb) bomb load

Lavochkin La-7

Unknown unit, central part of the Eastern Front, 1944–45

The La-7 was the final development of the already excellent La-5.

Specifications

Crew: 1
Powerplant: 1380kW (1850hp) Shvetsov M-82FN radial piston engine
Maximum speed: 680km/h (425mph)
Range: 990km (618 miles)
Service ceiling: 9500m (31,160ft)
Dimensions: span 9.80m (32ft 2in); length 8.60m (28ft 3in)
Weight: 3280kg (7231lb) loaded
Armament: 2 x 20mm (0.8in) ShVAK cannon or 3 x 20mm (0.8in) Berezin B-20 cannon in nose; plus up to 200kg (441lb) bomb load

Lavochkin La-7

163rd Guards IAP, Eastern Front, 1944–45

The La-7 was without doubt one of the finest fighters of World War II, and this example is believed to have been the personal aeroplane of the 163rd Guards IAP's commander.

the Finns in the north, leading to Finland's armistice in September. The Soviet summer offensive did not end with the liberation of Byelorussia, but continued remorselessly onwards. Even though they were approaching the point of exhaustion and their lines of communication were stretched to the very limit, the Red Army advanced into eastern Poland before finally halting outside Warsaw on 1 August.

Specifications

Crew: 1

Powerplant: 895kW (1200hp) Allison V-1710-85 liquid-cooled V-12

Maximum speed: 605km/h (376mph)

Range: 483km (300 miles)

Service ceiling: 11,665m (38,270ft)

Dimensions: span 10.36m (34ft 0in); length 9.195m (30ft 2in); height 3.785m (12ft 5in)

Weight: 3992kg (8800lb) loaded

Armament: 1 x 37mm (1.46in) M4 cannon, 2 x 12.7mm (0.5in) Browning MGs, 4 x 7.62mm (0.3in) Browning MGs

▲ **Bell P-39Q Airacobra**

Unknown unit, southern Soviet Union, 1944

This is another of the Bell P-39s flown by Major Aleksandr Pokryshkin, the second-highest scoring Soviet ace with a total of 59 victories. On 16 July 1944, Pokryshkin fought a group of Ju-87s and Hs-129s, shooting down three Ju-87s and one Hs-129.

Assault on the Reich

October 1944 – May 1945

The Soviets advanced into the Baltic states during September 1944, trapping a complete army group, and then through the Balkans right until the end of the year. By the start of 1945, they were poised for their advance through Germany, culminating in the capture of Berlin.

By the autumn of 1944, the V-VS was rampant over the Eastern Front. It prevented the rapidly declining German air strength from achieving anything of note and provided the Soviet ground forces with tactical support of unparalleled capability over the battlefield and against the German lines of communication.

Overwhelming odds

The German continued to fight hard on the ground, but were outnumbered and completely outgunned, and therefore could not halt the Soviet steamroller. The Soviets cleared the Baltic states, except for an isolated pocket in which a German army group survived to the end of the war to no good purpose, and at the same time moved through the Balkans to approach Yugoslavia and enter eastern Hungary.

On 12 January 1945, the Soviets launched a huge offensive between the Carpathian mountains in the south and the Baltic Sea in the north, while at the same time another thrust advanced through Hungary to reach Vienna on 15 April.

Push for Berlin

The prize most desired by the Soviets, however, was Berlin, and it was on 22 April that the Soviet advance guards reached the outskirts of the German capital, which had been surrounded by 25 April. The Battle of Berlin lasted until 2 May, three days after Hitler's suicide and just six days before Germany's unconditional surrender.

The Soviets estimated, probably correctly, that in the previous three months the Germans had lost about one million men killed, while Soviet captures amounted to 800,000 men, 6000 aircraft, 12,000 armoured fighting vehicles and 23,000 pieces of artillery. The Red Army and its air forces had delivered a final, terrible revenge upon Germany and its people.

Specifications

Crew: 1
Powerplant: 1380kW (1850hp) Shvetsov M-82FN radial piston engine
Maximum speed: 680km/h (425mph)
Range: 990km (618 miles)
Service ceiling: 9500m (31,160ft)
Dimensions: span 9.80m (32ft 2in); length 8.60m (28ft 3in)
Weight: 3280kg (7231lb) loaded
Armament: 2 x 20mm (0.8in) ShVAK cannon or 3 x 20mm (0.8in) Berezin B-20 cannon in nose; plus up to 200kg (441lb) bomb load

▲ Lavochkin La-7

Unknown unit, Eastern Front, 1944–45

With its powerful air-cooled radial engine and clean lines, the La-7 offered excellent performance.

▲ Lavochkin La-7

176th Guards IAP / 302nd IAD (Fighter Division), Eastern Front, 1944–45

This was one of the aircraft flown by Colonel Ivan Kozhedub, the Soviet Union's highest-scoring ace. Three times decorated as Hero of the Soviet Union, Kozhedub is credited with downing 62 enemy aircraft.

Specifications

Crew: 1
Powerplant: 1380kW (1850hp) Shvetsov M-82FN radial piston engine
Maximum speed: 680km/h (425mph)
Range: 990km (618 miles)
Service ceiling: 9500m (31,160ft)
Dimensions: span 9.80m (32ft 2in); length 8.60m (28ft 3in)
Weight: 3280kg (7231lb) loaded
Armament: 2 x 20mm (0.8in) ShVAK cannon or 3 x 20mm (0.8in) Berezin B-20 cannon in nose; plus up to 200kg (441lb) bomb load

Chapter 5

Other Allied Air Forces

While the French, British, US and Soviet fighter arms dominated the fighting in Europe during World War II, there were also a number of smaller air arms involved. First of these was the Polish Air Force, which fought gallantly and with some success in the war's first campaign against a German opponent tactically, technically and operationally superior. But smaller European air arms that should also be mentioned, not least because they provided the Allies with aircrews after the defeat of their countries, were Denmark, Norway, the Netherlands, Belgium, Yugoslavia and Greece. Finally, there was Brazil, which sent a small expeditionary force to support the Allied cause during the Italian campaign, in which there was also a small Italian Co-Belligerent Air Force.

Gull-winged fighter

The fighters of the PZL organization were the mainstays of the Polish fighter arm in the 1930s. These are P.11 aircraft, which by 1939 were of obsolescent design.

Polish Air Force

September 1939

The Polish Air Force was only just moving into the era of the 'modern' monoplane design in 1939, and was outclassed by a *Luftwaffe* equipped with more advanced aircraft and, just as importantly, using tactics proved in the Spanish Civil War. Even so, the Poles fought hard.

Most of Poland's first-line warplanes at the time of the German invasion of 1 September were products of the PZL organization, and of these the most advanced were the P.23 Karas single-engine and P.37 Los twin-engine aircraft operated in the reconnaissance and bomber roles.

The squadrons flying these types were aggressive in their efforts, but lacked the training and fighter cover to survive long in the face of the Germans' very capable fighter arm. From the middle of September, moreover, the Germans concentrated a considerable effort against the Polish airfields and soon deprived the Poles of the bases from which the surviving aircraft could operate.

ORGANIZATION

Polish Air Force
September 1939

- Dispositional Air Force
- Armies' Air Force

AIRCRAFT OF THE DISPOSITIONAL AIR FORCE'S PURSUIT BRIGADE

Unit	Type	Strength
III/1 (111 & 112 Squadrons)	PZL P.11	20
IV/1 (113 & 114 Squadrons)	PZL P.11	20
III/2 (123 Squadron)	PZL P.7a	10

Aggressive tactics

The Polish fighter arm was still more determined and aggressive in its tactics, but was poorly served in its aircraft, which were gull-winged fighters with fixed landing gear. The oldest of these in service in September 1939 was the P.7, which had entered service in 1932. The P.11 was conceptually very similar but with a more powerful engine, and claimed 125 victories in the air before the defeat of Poland.

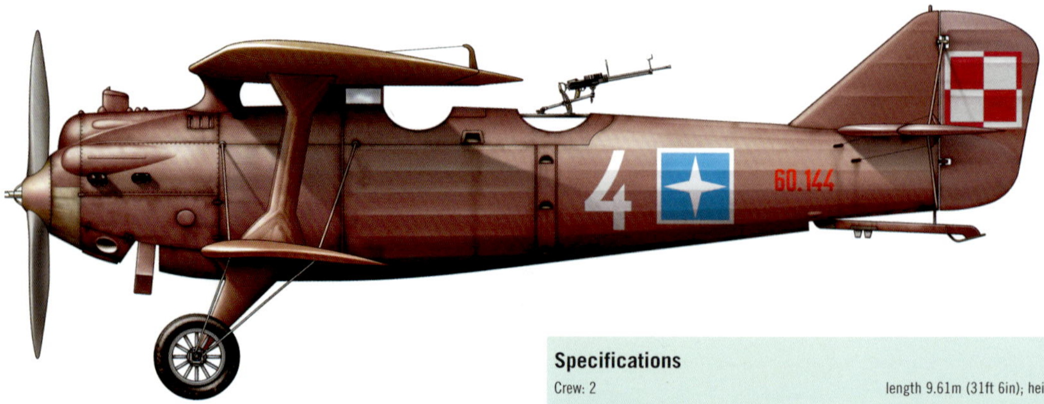

▲ **Breguet Bre.19**

Unknown unit, Polish Air Force, 1930s

An obsolete type, the Bre.19 remained in Polish service well into the 1930s, latterly in second-line roles.

Specifications

Crew: 2
Powerplant: 336kW (450hp) Lorraine 12Ed piston
Maximum speed: 214km/h (133mph)
Range: 800km (497 miles)
Service ceiling: 7200m (23,620ft)
Dimensions: span 14.83m (48ft 7in); length 9.61m (31ft 6in); height 3.69m (12ft 1in)
Weight: 1387kg (3058lb)
Armament: 1 x fixed forward 7.62mm (0.3in) Vickers MG, and two flexibly mounted rear 7.62mm (0.3in) Lewis MGs

Specifications
Crew: 1
Powerplant: 470kW (630hp) Bristol Mercury V S2 radial engine
Maximum speed: 375km/h (233mph)
Range: 550km (341 miles)
Service ceiling: 8000m (26,246ft)
Dimensions: span 10.72 m (35ft 2in); length 7.55m (24ft 9in); height 2.85m (9ft 4in)
Weight: 1650kg (3638lb) loaded
Armament: 2–4 7.92mm (0.312in) MGs; plus 50kg (110lb) bomb load

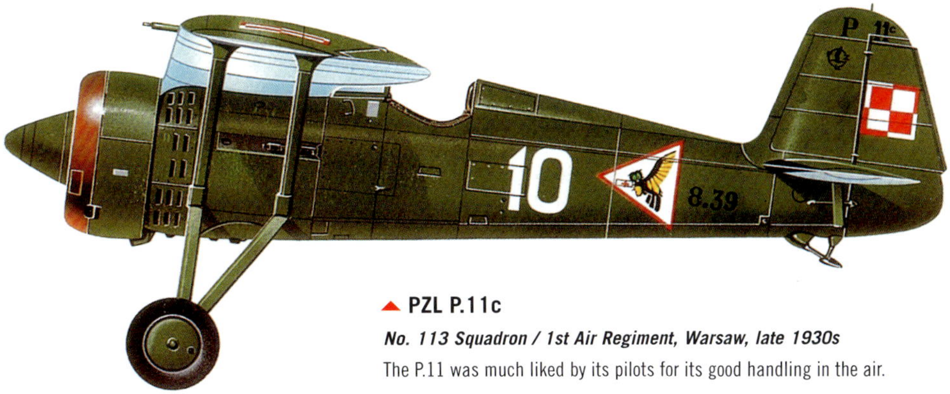

▲ PZL P.11c

No. 113 Squadron / 1st Air Regiment, Warsaw, late 1930s

The P.11 was much liked by its pilots for its good handling in the air.

▲ PZL P.11c

Unknown Polish unit, late 1930s

Although it had moderately good performance by the standards of the mid-1930s, the P-11 was obsolete at the start of World War II and easily outclassed by the airplanes of the *Luftwaffe*.

Specifications
Crew: 1
Powerplant: 470kW (630hp) Bristol Mercury V S2 radial engine
Maximum speed: 375km/h (233mph)
Range: 550km (341 miles)
Service ceiling: 8000m (26,246ft)
Dimensions: span 10.72 m (35ft 2in); length 7.55m (24ft 9in); height 2.85m (9ft 4in)
Weight: 1650kg (3638lb) loaded
Armament: 2–4 7.92mm (0.312in) MGs; plus 50kg (110lb) bomb load

Specifications
Crew: 1
Powerplant: 470kW (630hp) Bristol Mercury V S2 radial engine
Maximum speed: 375km/h (233mph)
Range: 550km (341 miles)
Service ceiling: 8000m (26,246ft)
Dimensions: span 10.72 m (35ft 2in); length 7.55m (24ft 9in); height 2.85m (9ft 4in)
Weight: 1650kg (3638lb) loaded
Armament: 2–4 7.92mm (0.312in) MGs; plus 50kg (110lb) bomb load

▲ PZL P.11c

No. 122 Squadron / 2nd Air Regiment, Krakow, late 1930s

This was numerically the most important fighter in Polish service at the time of the German invasion of September 1939.

Low Countries

May 1940

When Germany invaded the Netherlands and Belgium in May 1940, both of these countries had small air forces, but they lacked operational experience and the size to do anything but inflict modest losses on the Germans as their ground forces were driven into defeat.

THE FOKKER company had designed and built large numbers of capable warplanes, mainly for export, during the 1920s, but had then started to fall behind as larger nations began to introduce the 'modern' type of monoplane warplane with a cantilever wing and retractable main landing gear units.

Thus the country's most modern aircraft was the Fokker G.I twin-engine heavy fighter with adequate performance and heavy armament. However, only 23 were operational at the time of the German invasion of 10 May 1940, and these could be of little real use in the five-day campaign culminating in the Netherlands' defeat, by which time only one of the aircraft was left.

The most important Dutch fighter was the Fokker D.XXI, which may be categorized as an interim type since it had a low-set cantilever wing and enclosed cockpit, but retained fixed main landing gear units and an obsolescent metal-framed structure covered with fabric. The type had first flown in 1936, and by May 1940, 36 of the aircraft had been delivered, of which 29 were used in the five-day campaign. The Dutch pilots flew their aircraft with considerable courage, but most of the aircraft were destroyed either in the air or on the ground.

ORGANIZATION

Dutch Air Force
May 1940

1st Air Regiment | 2nd Air Regiment

FIGHTER UNITS OF THE DUTCH AIR FORCE

Unit	Type	Strength	Base
1st Fighter Sqn/1 Reg	Fokker D.XXI	11	de Kooy
2nd Fighter Sqn/1 Reg	Fokker D.XXI	9	Schiphol
3rd Fighter Sqn/1 Reg	Fokker G.IA	11	Waalhaven
4th Fighter Sqn/1 Reg	Fokker G.IA	12	Bergen -op-Zoom
1st Fighter Sqn/2 Reg	Fokker D.XXI	8	Ypenberg
3rd Fighter Sqn/2 Reg	Douglas DB-8A	11	Ypenberg

Fokker D.XXI

Dutch Air Force, 1939–40

The D.XXI was an interim monoplane with a cantilever wing and enclosed cockpit, but had an obsolete structure and fixed landing gear. When it entered service in 1938, it was a quantum leap forward for the Dutch Army Aviation Group. Until then, its fighter force had consisted of ageing biplanes with open cockpits. The new Fokker proved to be an extremely sturdy aircraft.

Specifications

Crew: 1
Powerplant: 619kW (830hp) Bristol Mercury VIII air-cooled, 9-cylinder, radial
Maximum speed: 418km/h (260mph)
Range: 930km (574 miles)
Service ceiling: 9350m (30,675ft)
Dimensions: span 11m (36ft 1in); length 8.20m (26ft 11in); height 2.95m (9ft 8in)
Weight: 1970kg (4399lb) maximum takeoff
Armament: 4 x 7.92mm (0.34in) FN Browning M36 MGs

▲ **Hawker Hurricane Mk I**

2e Escadrille / 2e Regiment de l'Aeronautique / Belgian Air Force, Diest Schaffen, 1940

Some 11 of the 18 Hurricane aircraft that Belgium received were in service at the time of the German invasion in May 1940.

Specifications

Crew: 1
Powerplant: 768kW (1030hp) Rolls-Royce Merlin II liquid-cooled V-12
Maximum speed: 496km/h (308mph)
Range: 845km (525 miles)
Service ceiling: 10,180m (33,400ft)
Dimensions: span 12.19m (40ft 0in); length 9.55m (31ft 4in); height 4.07m (13ft 4.5in)
Weight: 2820kg (6218lb) loaded
Armament: 8 x 7.7mm (0.303in) Browning MGs

▲ **Hawker Hurricane Mk IA**

2e Escadrille / 2e Regiment de l'Aeronautique / Belgian Air Force, Diest Schaffen, 1940

Belgium ordered 20 British-built Hurricane Mk IAs and 80 for local construction, but received 15 and built three respectively.

Specifications

Crew: 1
Powerplant: 768kW (1030hp) Rolls-Royce Merlin II liquid-cooled V-12
Maximum speed: 496km/h (308mph)
Range: 845km (525 miles)
Service ceiling: 10,180m (33,400ft)
Dimensions: span 12.19m (40ft 0in); length 9.55m (31ft 4in); height 4.07m (13ft 4.5in)
Weight: 2820kg (6218lb) loaded
Armament: 8 x 7.7mm (0.303in) Browning MGs

Belgian Air Force

With only a small indigenous aircraft manufacturing industry and limited financial resources, Belgium opted to concentrate on the local manufacture or assembly of British aircraft by the Belgian subsidiary of the Fairey company of the UK.

The air force also flew 23 examples of a late-generation Italian biplane fighter, the Fiat CR.42 Falco, derived from the CR.32 that had performed creditably in the Spanish Civil War. The CR.42 was wholly outclassed by the Germans' Messerschmitt Bf 109 monoplane fighter. Much the same can be said of

ORGANIZATION

Belgian Air Force
May 1940

1st Air Regiment (six army co-operation squadrons)	2nd Air Regiment (six fighter squadrons)	3rd Air Regiment (four reconnaissance bomber squadrons)

the Gloster Gladiator, the last British biplane fighter to enter service. At the time of the German invasion, the Belgians had 15 of these fighters in service contained in one squadron.

Without doubt the best fighter in Belgian service was another British type, the Hawker Hurricane, of which there were 11 in service with more on order. The Hurricane was a capable blend of obsolescent and modern, the former represented by its metal-framed and largely fabric-covered airframe, and the latter by its overall configuration, retractable main landing gear units, enclosed cockpit, and powerful fixed forward-firing armament of eight rifle-calibre machine guns in the leading edges of its wing.

This mix of fighters could avail the Belgian Air Force little in May 1940 before inevitable defeat.

▲ **Gloster Gladiator Mk I**

1e Escadrille 'La Comete' / 2e Regiment de l'Aeronautique / Belgian Air Force, Diest Schaffen, 1940

The Belgians received 22 Gladiators, 15 of which were operational in May 1940.

Specifications

Crew: 1

Powerplant: 619kW (830hp) Bristol Mercury VIIIAS air-cooled 9-cylinder radial

Maximum speed: 407km/h (253mph)

Range: 684km (425 miles)

Service ceiling: 9845m (32,300ft)

Dimensions: span 9.83m (32ft 3in); length 8.36m (27ft 5in); height 3.52m (11ft 7in)

Weight: 2272kg (5020lb) loaded

Armament: 4 x 7.7mm (0.303in) Browning MGs

Scandinavia

April–May 1940

Before turning its attention to the west, where the Netherlands and Belgium were defeated as part of the great German offensive against France from 10 May, Germany decided to secure its northern flank by the swift overrunning of Denmark and the defeat of Norway.

Small in size and and with only the most limited of armed forces, Denmark could offer no realistic opposition to the German invasion of 9 April 1940, which was intended to provide support for the more important conquest of Norway. Denmark's air force was miniscule and wholly obsolete in its aircraft, the most modern type being the Gloster Gladiator that replaced the Bristol Bulldog in one squadron from 1935. Morevoer, it saw no combat since Denmark swiftly capitulated in return for the German offer of limited independence in internal matters.

Norwegian Air Force

Reflecting the country's larger size and greater population, the defence establishment of Norway was larger than that of Denmark. The army and navy each had their own air arms, both of which had operated the indigenously designed Hover MF.9 fighter, which was out of service by 1940. By this time, the army and naval air arms had establishments that permitted 36 and 20 fighters respectively, and the fighters in service were the Curtiss Hawk 75A first-generation 'modern' monoplane from the United States and the Gloster Gladiator last-generation biplane from the United Kingdom.

In these circumstances, therefore, the Norwegian air arms could offer little effective resistance to the German invasion. The Norwegian pilots flew with courage and determination, but their lack of operational experience and the obsolescence of their equipment made the result a foregone conclusion. The Norwegian losses were comparatively heavy but, like the Poles before them, the Norwegians flew some of their aircraft to safety, in this instance to the UK. Numbers of Norwegian personnel escaped to carry on the fight from the United Kingdom and, in the case of naval personnel, Canada.

▲ **Bristol Bulldog Mk IIA**

Danish Air Force, 1939

Denmark received four of these aircraft with unsupercharged engines.

Specifications

Crew: 1
Powerplant: 490kW (365hp) Bristol Jupiter VII radial piston engine
Maximum speed: 287km/h (178mph)
Range: n/a
Service ceiling: 8930m (29,300ft)
Dimensions: span 10.3m (33ft 10in); length 7.67m (25ft 2in); height 2.67m (8ft 9in)
Weight: 1586kg (3490lb) loaded
Armament: 2 x 7.7mm (0.303in) Vickers MGs; Four x 9kg (20lb) bombs

▼ **Gloster Gladiator Mk I**

Norwegian Air Force, 1939

The Norwegian air force received 12 Gladiator fighters in the form of six Mk I and six Mk II aircraft, and these were entrusted with the air defence of Oslo.

Specifications

Crew: 1
Powerplant: 619kW (830hp) Bristol Mercury VIIIAS air-cooled 9-cylinder radial
Maximum speed: 407km/h (253mph)
Range: 684km (425 miles)
Service ceiling: 9845m (32,300ft)
Dimensions: span 9.83m (32ft 3in); length 8.36m (27ft 5in); height 3.52m (11ft 7in)
Weight: 2272kg (5020lb) loaded
Armament: 4 x 7.7mm (0.303in) Browning MGs

Yugoslavia and Greece

April 1941

Before turning the German forces against the Soviet Union, Hitler decided to secure his southern flank by seizing Yugoslavia, which had turned against the German-Italian axis, and Greece, from which Allied bombers might be able to cripple the Romanian oilfields on which Germany relied.

Although beset by internal divisions and a sadly deficient infrastructure, the Yugoslav forces were large and, on paper at least, comparatively formidable at the time of the German invasion of 6 April 1941. The Yugoslav Air Service reached a peak in 1935, when it had 44 squadrons with 440 aircraft, most of

▲ **Breguet Bre 19 A2**

Royal Yugoslav Air Force, Kraljevo, 1936

The Bre 19 A2 was obsolete by the time of the German invasion, and was used mainly for reconnaissance purposes. After the fall of the Kingdom of Yugoslavia, 46 aircraft were seized and used by the Croatian Air Force for anti-partisan missions.

Specifications

Crew: 2
Powerplant: 336kW (450hp) Lorraine 12Ed piston
Maximum speed: 214km/h (133mph)
Range: 800km (497 miles)
Service ceiling: 7200m (23,620ft)
Dimensions: span 14.83m (48ft 7in); length 9.61m (31ft 6in); height 3.69m (12ft 1in)
Weight: 1387kg (3058lb)
Armament: 1 x fixed forward 7.62mm (0.3in) Vickers MG, and mountings for two rear 7.62mm (0.3in) Lewis MGs

▲ **Hawker Hurricane Mk I**

Royal Yugoslav Air Force, Zemun, April 1941

This is one of 20 Hurricanes built at the factory at Zmaj.

Specifications

Crew: 1
Powerplant: 768kW (1030hp) Rolls-Royce Merlin II liquid-cooled V-12
Maximum speed: 496km/h (308mph)
Range: 845km (525 miles)
Service ceiling: 10,180m (33,400ft)
Dimensions: span 12.19m (40ft 0in); length 9.55m (31ft 4in); height 4.07m (13ft 4.5in)
Weight: 2820kg (6218lb) loaded
Armament: 8 x 7.7mm (0.303in) Browning MGs

them obsolescent if not obsolete. Hawker Furys and Breguet Bre 19s made up the bulk of the fighter aircraft available.

A major re-equipment programme had been launched, and this included deliveries of the Curtiss P-40B, Hawker Hurricane Mk I and Messerschmitt Bf 109E fighters, as well as the start of licensed production of the Hurricane.

Local effort was responsible for two other fighter types, the Ikarus IK-2 gull-wing monoplane with fixed landing gear, and the more advanced Rogozarski IK-3 monoplane with retractable landing gear. The former entered service in 1939, and eight of the 12 such aircraft were operational in April 1941. The latter entered service in 1940, but only six were operational at the time of the invasion.

Greece

Greece was invaded at the same time as Yugoslavia, and its fighter arm at this time operated the Bloch MB.151 and PZL P.24 fighters from France and Poland. Neither could compete effectively with the German fighters they faced.

▲ Monoplane wings over Yugoslavia

The Yugoslav Air Force included more than 30 operational Hurricane Mk Is at the time of the German invasion.

▼ Hawker Fury

35th Fighter Group / 109 Eskadrila / Royal Yugoslav Air Force, April 1941

The 35th and 36th Fighter Groups had 30 operational Furys at the time of the German invasion. Yugoslav Furies did see action against Axis forces. On 6 April 1941, a squadron of Furies engaged with some Messerschmitt Bf 109Es and Messerschmitt Bf 110s. In the resulting conflict, 11 Furies were destroyed – almost the entire squadron – but seven German aircraft also failed to return, though it is possible that some of these were non-combat losses. However, it is certain that at least one of them was lost to ramming by a Fury.

Specifications

Crew: 1

Powerplant: 477kW (640hp) Rolls-Royce Kestrel IV V12 engine

Maximum speed: 360km/h (223mph)

Range: 435km (270 miles)

Service ceiling: 8990m (29,500ft)

Dimensions: span 9.14 m (30ft); length 8.15m (26ft 9in); height 3.10m (10ft 2in)

Weight: 1637kg (3609lb) loaded

Armament: 2 x 7.7 mm (0.303in) Vickers Mk IV MGs

Co-Belligerent Italy and Brazil

1943–45

When Italy agreed an armistice with the Allies in September 1943, the northern half of the country was still in German hands but the southern portion worked actively with the Allies. Another Allied supporter was Brazil, which sent a small expeditionary force to Italy.

On 9 September 1943, as the Allies landed on the Italian mainland just south of Naples, it was revealed that the Italian government had secured an armistice with the Allies, and many Italian squadrons flew south as and when they could to the safety of the Allied-held airfields in the south of Italy. At the same time, the Germans occupied the whole of northern Italy, where a revived fascist state was established under Benito Mussolini, the Italian dictator deposed in July of the same year.

Italian Co-Belligerent Air Force

The forces of this revived state continued to fight alongside the Germans to increasingly little effect, while the Allies permitted the establishment of an Italian Co-Belligerent Air Force (*Aviazione Cobelligerante Italiana*) in southern Italy during October 1943 to fly alongside their own squadrons. That part of the previous Regia Aeronautica now under German control was the National Republican Air Force (*Aeronautica Nazionale Repubblicana*), which was nominally part of the forces of the Italian Social Republic.

By the end of 1943, some 281 Italian warplanes had landed on Allied airfields, but most of these were no longer useful for combat, especially as production was centred in German-held northern Italy and spares were therefore very short. The Italian units were re-equipped with Allied warplanes, and were thereafter employed in tasks such as transport, escort, reconnaissance, sea rescue and limited tactical ground support operations. The Co-Belligerent Air Force flew something in the order of 11,000 missions from

ORGANIZATION

Italian Co-Belligerent Air Force
Late 1944

- 28o Gruppo, Stormo Baltimore
- 2o Gruppo, 30 Stormo Transporto

▲ **Bell P-39Q Airacobra**

Italian Co-Belligerent Air Force, 1944

Italy received 149 examples of the Airacobra from US Fifteenth Air Force stocks, and these saw limited operational use in World War II.

Specifications

Crew: 1
Powerplant: 895kW (1200hp) Allison V-1710-85 liquid-cooled V-12
Maximum speed: 605km/h (376mph)
Range: 483km (300 miles)
Service ceiling: 11,665m (38,270ft)
Dimensions: span 10.36m (34ft 0in); length 9.195m (30ft 2in); height 3.785m (12ft 5in)
Weight: 3992kg (8800lb) loaded
Armament: 1 x 37mm (1.46in) M4 cannon, 2 x 12.7mm (0.5in) Browning MGs, 4 x 7.62mm (0.3in) Browning MGs, and 227kg (500lb) of bombs

October 1943 to April 1945. None of its missions was flown over Italy to avoid any possibility of a civil war, the missions instead being flown over the Balkans. It formed the basis for the post-war Italian Air Force, the *Aeronautica Militare Italiana*.

Brazilian flyers

Brazil declared war on Germany in 1942, and in 1943 the 1st Air Fighter Unit was formed from volunteers for service in Italy. The group had four 12-aircraft squadrons, and totalled 350 men, including 43 pilots. After combat training under US supervision in Panama, the group was moved to the United States, transitioned to the Republic P-47D Thunderbolt heavy fighter, and was shipped to Italy, where it arrived in October 1944. The pilots were allocated to the US 350th Fighter Group, in which the Brazilian unit was the 1st Fighter Squadron. This unit flew 2546 sorties in 445 missions.

Carrying the moto 'Senta a Pua' (meaning 'hit 'em hard'), the badge of the Brazilian 1o Gruppo de Caca was designed by the pilots of the unit while in transit to Europe. The green-yellow surrounding represents Brazil, while the red field represents the war skies; the white clouds represent the ground and the blue shield charged with the Southern Cross is the common symbol for the Brazilian armed forces.

Specifications

Crew: 1
Powerplant: 1891kW (2535hp) Pratt & Whitney R-2800-59W Double Wasp
Maximum speed: 697km/h (433mph)
Range: 3060km (1900 miles) with drop tanks
Service ceiling: 12,495m (41,000ft)
Dimensions: span 12.42m (40ft 9in); length 11.02m (36ft 2in); height 4.47m (14ft 8in)
Weight: 7938kg (17,500lb) maximum takeoff
Armament: 8 x 12.7mm (0.5in) MGs in wings, plus provision for 1134kg (2500lb) external bombs or rockets

▲ Republic P-47D Thunderbolt

1o Gruppo de Caca / Brazilian Air Force, Tarquinia, Italy, November 1944

The Brazilian Air Force received P-47Ds, all of which fought in Italy.

▲ Republic P-47D Thunderbolt

1o Gruppo de Caca / Brazilian Air Force, Tarquinia, Italy, November 1944

The 1°GAVCA clocked up 5465 combat flight hours from 11 November 1944 to 4 May 1945, mostly in the ground attack role, destroying bridges, ammunition depots and enemy vehicles.

Specifications

Crew: 1
Powerplant: 1891kW (2535hp) Pratt & Whitney R-2800-59W Double Wasp
Maximum speed: 697km/h (433mph)
Range: 3060km (1900 miles) with drop tanks
Service ceiling: 12,495m (41,000ft)
Dimensions: span 12.42m (40ft 9in); length 11.02m (36ft 2in); height 4.47m (14ft 8in)
Weight: 7938kg (17,500lb) maximum takeoff
Armament: 8 x 12.7mm (0.5in) MGs in wings, plus provision for 1134kg (2500lb) external bombs or rockets

Appendices

Neither Germany nor Japan expected or planned for long wars of attrition when they launched their attacks on Poland in 1939 and at Pearl Harbor in 1941. The miscalculation was to cost them dearly when Allied industrial capabilities rapidly increased from late 1942.

▼ Combat strengths, all countries

World War II saw sophisticated weaponry being created in unprecedented numbers. Nowhere was that more true than in the air, with tens of thousands of combat aircraft being deployed in the later years of the war. (Rounded figures are estimates.)

FRONTLINE COMBAT AIRCRAFT, COMPARATIVE STRENGTHS (1939–45)

Date	Germany	USA	USSR	UK	Total Allied
September 1939	2,916	–	–	1,660	1,660
August 1940	3,015	–	–	2,913	2,913
December 1940	2,885	–	–	1,064	1,064
June 1941	3,451	–	8,105	3,106	11,211
December 1941	2,561	4,000	2,495	4,287	6,782
June 1942	3,573	1,902	3,160	4,500	9,562
December 1942	3,440	4,695	3,088	5,257	13,040
June 1943	5,003	8,586	8,290	6,026	22,902
December 1943	4,667	11,917	8,500	6,646	27,063
June 1944	4,637	19,342	11,800	8,339	39,481
December 1944	5,041	19,892	14,500	8,395	42,787
April 1945	2,175	21,572	17,000	8,000	46,752

▼ Yak-9 fighters

The inscription on the side of these Yak-9s, 'Little Theatre: Front', indicates that they were donated by Moscow's Little Theatre in 1943.

German aircraft losses by type

Few German pilots who took part in the *Luftwaffe*'s early victories in Poland, France, the Balkans and the Soviet Union could have predicted the unsustainable losses they would suffer later in the conflict.

GERMAN AIRCRAFT LOSSES BY TYPE (1939–45)

Type	Total
Fighter	41,452
Night fighter	10,221
Ground attack	8,548
Bombers	22,037
Transport	6,141
Reconnaissance	6,733
Trainers	15,428

USAAF aircraft losses

The contrast between the Pacific and European theatres was at its height in 1944, where the Germans were destroying nine American aircraft for every one lost to the Japanese.

US AIRCRAFT OPERATIONAL LOSSES (1942–45)

Year	Europe	Pacific	Total
1942	141	344	485
1943	3,028	819	3,847
1944	11,618	1,671	13,289
1945	3,631	1,699	5,330
Total	18,418	4,533	22,951

Soviet aircraft losses

Soviet Air Forces suffered dreadfully in the early stage of the war. The consistent quality of the *Luftwaffe* opposition meant that loss rates remained high until the end of the conflict.

USSR COMBAT AIRCRAFT LOSSES (1941–45)

Period	Total
21 June – 30 Nov 1941	12,652
1 Dec – 30 April 1942	7,099
1 May – 31 Oct 1942	14,601
1 Nov – 30 June 1943	17,690
1 July – 31 Dec 1943	20,741
1 Jan – 31 May 1944	13,386
1 June – 31 Dec 1944	20,283
1 Jan – 8 May 1945	?
Total	106,652

Polish aircraft losses

The first *Blitzkrieg* campaign saw much of the Polish Air Force destroyed on the ground, and although those aircraft that did get into the air fought hard, they were no match for the *Luftwaffe*. By the end of September the few remaining operational aircraft had fled to Romania.

POLISH AIRCRAFT LOSSES (SEPTEMBER 1939)

Type	Total
Fighter	116
Dive-bomber	112
Bombers	36
Reconnaissance	81
Transport	9
Seaplane	21
Other	23

Aircraft losses, totals for all countries

Combat losses accounted for only part of the totals of aircraft destroyed in World War II. Operational service was hard, and rough fields, bad weather and mechanical failures accounted for almost as many aircraft as the enemy.

ALL OPERATIONAL LOSSES (1939–45)

Country	Lost in action	Damaged	Total
Finland	251	352	603
France	413	479	892
Germany	69,583	47,001	116,584
Italy	3,269	2003	5,272
Japan	–	–	38,105
Netherlands	–	–	81
Poland	–	–	398
UK	–	–	22,010
USA	–	–	22,951
USSR	–	–	106,652

RAF fighter strengths by theatre

Early in the war the bulk of RAF fighter strength was reserved for the defence of the home islands. As the war expanded worldwide, fighting units saw action in increasing numbers in all of the major theatres.

RAF COMBAT FIGHTER SQUADRONS (1939–45)

Date	UK	NW Europe	Med	Far East	Total
September 1939	41	4	6	4	55
July 1940	65	–	8	4	77
December 1941	114	–	29	13	156
December 1942	93	–	47	19	159
December 1943	83	–	49	29	161
September 1944	58	48	41	29	176
March 1945	46	51	33	30	160

FAA fighter/fighter-bomber strengths by theatre

The Fleet Air Arm was very much a poor relation of the RAF at the outbreak of war, but it was to grow twenty-fold over the course of the conflict to play an important part in maritime warfare.

FLEET AIR ARM FIGHTER SQUADRONS (1939–45)

Date	Total	Far East	Aircraft
September 1939	3	–	36
September 1940	7	–	78
September 1941	13	–	129
September 1942	27	–	252
September 1943	32	–	339
September 1944	37	4	645
April 1945	39	26	826
September 1945	40	32	739

Eastern Front, combat aircraft

The onset of Operation *Barbarossa* in the summer of 1941 saw the Soviet air forces reduced to parity with the *Luftwaffe* by the time of the battle for Moscow, but Soviet production had outstripped the Germans within a year.

SOVIET AND GERMAN FRONT-LINE COMBAT AIRCRAFT (1941–45)

Date	German	Soviet
June 1941	2,130	8,100
July 1941	1,050	2,500
December 1941	2,500	2,500
May 1942	3,400	3,160
November 1942	2,450	3,100
July 1943	2,500	8,300
January 1944	1,800	8,500
June 1944	1,710	11,800
January 1945	1,430	14,500
April 1945	1,500	17,000

USAAF fighter combat groups by theatre

In US usage, a combat group was a formation consisting of two or more squadrons, with two or more groups forming a wing. By far the largest proportion of American combat strength during World War II was deployed to Europe and the Mediterranean, though large numbers were due to be transferred to the Pacific when the Japanese surrendered in 1945.

USAAF FIGHTER COMBAT GROUPS STATIONED OVERSEAS BY THEATRE (1941–45)

Date	Pacific	CBI	UK	NW Europe	Med	Total
December 1941	4	–	–	–	–	4
December 1942	7	2	4	–	10	23
December 1943	10	4	17	–	13	44
September 1944	9	6	16	17	12	60
March 1945	12	6	9	21	10	58

▲ **Spitfire Mk VC**

Serving with the Desert Air Force (DAF) in 1943, this Spitfire of No. 43 Squadron is parked at a captured Italian airfield at Comiso, Sicily.

▼ **Fighter production, all Allied and Axis forces**

In terms of economic output, the Axis powers were no match for the Allies, with the United States alone producing more aircraft than Germany, Italy and Japan combined. Albert Speer's miraculous industrial reorganization in the face of heavy Allied bombing increased German production fivefold between 1942 and 1944, but even this could not match the combined industrial might of the USA, the Soviet Union and the British Empire.

ANNUAL ALLIED MILITARY AIRCRAFT PRODUCTION (1939–45)

Year	USA	USSR	UK	Total
1939	–	–	1,324	1,324
1940	1,162	4,574	4,283	10,019
1941	4,416	7,086	7,064	18,566
1942	10,769	9,924	9,849	30,542
1943	23,988	14,590	10,727	49,305
1944	38,873	17,913	10,730	67,516
1945	20,742	c 9,000	5,445	35,187
TOTAL	99,950	63,087	49,422	212,459

ANNUAL AXIS MILITARY AIRCRAFT PRODUCTION (1939–45)

Year	Germany	Italy	Japan	Total
1939	605	?	?	?
1940	2,746	1,155	?	?
1941	3,744	1,339	1,080	6,163
1942	5,515	1,488	2,935	9,938
1943	10,898	528	7,147	18,573
1944	26,326	–	13,811	40,137
1945	5,883	–	5,474	11,357
TOTAL	55,727	4,510	30,447	90,684

Index

Page numbers in *italics* refer to illustrations.